multiplication and division games and ideas

Gary G. Bitter, Ph. D.
  Arizona State University
  Tempe, Arizona

Jerald Mikesell, Ed. D.
  Mesa Schools
  Mesa, Arizona

Kathryn Maurdeff
  Mesa Schools
  Mesa, Arizona

INSTRUCTO/McGraw-Hill
Paoli, Pennsylvania 19301

Editorial Staff: Research and Development Department
Instructo/McGraw-Hill

Cover Design: Gilbert Lieberman

Illustrations: Carole Smith

Art Assistant: Angela Tocci

Library of Congress Cataloging in Publication Data

Bitter, Gary G
    Multiplication and division games and ideas.

    1. Multiplication--Study and teaching. 2. Division--Study and teaching. I. Mikesell, Jerald L., 1935-   joint author. II. Maurdeff, Kathryn, joint author. III. Title.
QA115.B57   372.7'2   75-25931
ISBN 0-07-082047-3

Table of Contents

How to Use this Book!

The book, "Why Johnny Can't Add," by Morris Kline has created a great interest among parents and teachers in children's mastery of the basic facts of arithmetic. With this in mind, "Multiplication and Division Games and Ideas," as well as "Addition and Subtraction Games and Ideas" were written to assist teachers and parents in providing activities for their students and children. It is hoped that the games and ideas in these two books will provide understanding and skill development in the basic facts of arithmetic.

A variety of objectives have been listed for each of the two basic skill areas included in this book. Games and ideas have been written to help students meet each of these objectives. By determining which objective a student has not mastered, teachers and parents will have a group of activities and ideas to remedy the diagnosed weaknesses.

Suggested Procedures:

1. Administer a basic skills diagnostic test to each student. You will find samples of a Multiplication Test and a Division Test in Appendixes A and B. We suggest that you use tests in which you have confidence to determine if a student has achieved mastery of the listed objectives.

2. Record the student's progress on each skill on a profile form, similar to the ones in Appendix B. A possible coding procedure would be:

   ✓ has little or no understanding;

   ⊘ has some understanding, but lacks mastery;

   ● answered problem correctly.

   By referring to a profile chart you can plan group activities or interest centers that will meet the specific needs of the students and make their learning more meaningful and enjoyable.

Answers for the activities are provided for the convenience of the teacher and can be placed on solution activity cards for the interest center format. Space has been provided throughout this book for personal notations by the user. Hopefully, then, this book will provide some alternatives which are fun and challenging for the students who need more understanding and practice in the basic skills.

Good Luck!

Multiplication Objectives

Picture Perfect

Grade Levels: 2 - 3

Number of Participants: any number

Materials Required: Counters (bottle caps, straws, buttons, etc.)

Directions:

1. Pass out 10 counters or less to the participants.

2. Have participants arrange their counters as shown or in any number of rows and columns.

3. Ask the following questions with a point being given to the first player who answers each question correctly.

   a. How many objects in each row?

   b. How many in each column?

   c. How many total?

4. The player with the most points at the end of the game is the winner.

IDEA: Arrange objects as illustrated above on an overhead projector. Turn it on for just a few seconds, and then ask the same questions. This time, however, do not pass out any counters.

Inspector General

Grade Levels: 2 - 3

Number of Participants: any number

Materials Required: Pin on badges saying "Inspector General," one package per participant containing several objects (paper bag with counters would do).

Directions:

1. Package 2 - 4 objects together.

2. Pass out one package to a student.

3. Have players open their packages and determine the number of objects in the package they received.

4. If the player can name the number of objects she or he would have if he or she had 2, 3, or 4 packages just like the one the player received, he or she receives a badge, "Inspector General."

IDEA: Ask students how many bags they would need to have 6, 8, 9, or 10 objects.

Terrific Trader

Grade Levels: 2 - 3

Number of Participants: any number

Materials Required: Buttons, bottle caps, or straws (Terrific Trader has one color or type, the players another color or type.)

Directions:

1. One student is chosen to be the Terrific Trader.

2. The Terrific Trader distributes 24 bottle caps to each participant.

3. The Terrific Trader holds up 2 - 4 bottle caps, stating the value of each.

4. The first participant who can name the correct number of bottle caps needed to make a proper trade makes the trade with the Terrific Trader.

   Example: Terrific Trader holds up 3 bottle caps stating that each has a value of 2. The first player to state that he or she could exchange 6 of his or her caps for the Terrific Trader's would be the player to do the trading.

5. The first player who uses all of his or her original caps becomes the new Terrific Trader.

Multiplication - 2

Given a picture of a repeated set of objects (less than 12), the pupil will be able to choose the correct repeated addition expression from a list of such expressions.

Clap and Snap

Grade Levels: 2 - 3

Number of Participants: any number

Materials Required: None

Directions:

1. Select one person to leave the room.

2. While that person is out of the room, the class decides which repeated addition expression they will illustrate by repeated clapping - using a snap of the fingers to illustrate the "plus" sign.

3. The person returns to the class and tries to name the repeated addition expression with its sums.

   Example: Class does: "Clap, clap, clap, snap,
   Clap, clap, clap, snap,
   Clap, clap, clap".

   Person responds: $3 + 3 + 3 = 9$

4. If the person is successful in naming the repeated addition expression, he or she chooses a new person to leave the room.

5. If he or she is unsuccessful, that person leaves the room and tries again.

Fishy Sets

Grade Levels: 2 - 3

Number of Participants: 2 - 5

Materials Required: Fishing pole with magnet, pictures of repeated sets of objects on fish shaped construction paper (paper clip attached) box or container to hold fish, paper and pencil.

Directions:

1. Put paper fish in container.

2. The first player "fishes" with his or her pole until a fish is "caught".

3. To keep the fish the player must write the repeated addition expression and the sum correctly.

4. Players take turns.

5. The player with the most fish when the "pond" is empty is the winner.

IDEA: Each player (with a full pond of fish) continues to "fish" until he or she makes an error in writing the repeated addition expression. The object is to see if the player can empty the pond without missing.

Pupil Scramble

Grade Levels: 2 - 3

Number of Participants: any number

Materials Required: Cards with repeated addition expressions written on them.

Directions:

1. Divide the players into at least 2 groups.

2. Shuffle the cards and place them face down.

3. One member of each group leaves the room.

4. The first card is turned over and the remaining players in each group arrange themselves in groups to match the repeated addition expression on the card.

5. The members return to the room and each names the repeated addition expression his or her team members are showing.

6. If he or she fails, that person tries again. If successful, then that person will choose the next player to leave the room.

Mental Maze

Grade Levels: 2 - 6

Number of Participants: any number

Materials Required: None

Directions:

1. The leader asks participants to start with some named number.

2. The participants will then be asked to follow a sequence of instructions involving addition and subtraction of random numbers.

3. Frequently insert multiplying by one and zero in the group of directions.

   Example: Start with 4, add 2, multiply by 1, add 3, take away 6, multiply by zero, add 24. What number do you have?

IDEA: As the students are playing the above game ask them to illustrate one times various numbers and zero times several numbers.

NOTE: The type of instructions will vary according to grade level and ability of group.

Active Factor

Grade Levels: 3 - 5

Number of Participants: any number

Materials Required: None

Directions:

1. One pupil is chosen to be the "Active Factor".

2. The teacher will assign the "Active Factor" a number he or she will always multiply by. (Assign the numbers one and zero frequently).

3. Other pupils will take turns naming one factor and a product utilizing that factor; for example, pupil would say factor 3 with product of 24.

4. The "Active Factor" would take the factor and multiply her or his own number by the factor.

5. If the product of the "Active Factor" number and the factor given is higher than the product given by the first player, the "Active Factor" will indicate this by pointing a thumb up.

6. If the product of the "Active Factor" is lower than the product given by the pupil, she or he would indicate this by pointing a thumb down.

7. Other pupils in the group take turns naming factors and products until someone can guess the number the "Active Factor" is multiplying by.

8. The pupil who guesses the number becomes the new "Active Factor".

Connect 'Em

Grade Levels: 3 - 4

Number of Participants: any number

Materials Required: Spirit duplicating master

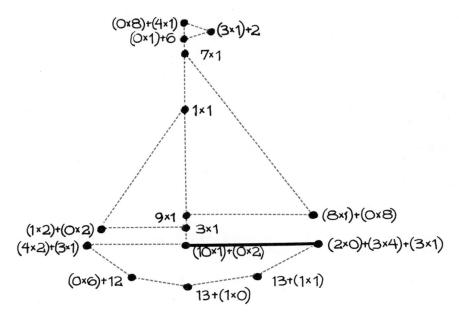

Directions:

1. Instead of a simple 1 - 2 - 3 dot-to-dot picture puzzle, the student must multiply the numbers mentally or multiply and then add the numbers mentally.

2. After doing the necessary computations, the pupil will draw the lines from 1 - 2 - 3 dot-to-dot.

3. Upon reaching number 15 the pupil will have a picture he or she will be able to recognize.

Secret Messages

Grade Levels: 3 - 5

Number of Participants: any number

Materials Required: Spirit duplicating master

Directions:

1. Do all of the problems given below.

2. Match your answers to the letters. The numbers under the message are the numbers of the problems.

   Example: The product for number 1 is 16. On page 13, the letter M has the value of 16 so put M in the blank above the number 1.

1. 4 x 4 = *16*	8. 8 x 4 = *32*	15. 3 x 3 = *9*
2. 7 x 2 = *14*	9. 3 x 4 = *12*	16. 5 x 8 = *40*
3. 5 x 7 = *35*	10. 0 x 8 = *0*	17. 3 x 3 x 3 = *27*
4. 4 x 5 = *20*	11. 3 x 5 = *15*	18. 2 x 2 x 2 = *8*
5. 3 x 8 = *24*	12. 4 x 7 = *28*	19. 3 x 3 x 5 = *45*
6. 7 x 3 = *21*	13. 5 x 6 = *30*	20. 2 x 5 = *10*
7. 5 x 1 = *5*	14. 6 x 7 = *42*	21. 9 x 2 = *18*

CODE:

A = 14	F = 18	M = 16	S = 21
B = 9	G = 27	N = 12	T = 35
C = 28	H = 20	O = 15	U = 32
D = 8	I = 24	P = 42	V = 10
E = 0	L = 30	R = 45	W = 40
			Y = 5

W   H   A   T   D   O   Y   O   U   R
16   4   2   3   18   11   7   11   8   19

S   H   O   E   S   B   E   C   O   M   F
6   4   11   10   6   15   10   12   11   1   10

W   H   E   N   Y   O   U   S   T   E   P
16   4   10   9   7   11   8   6   3   10   14

I   N   T   O   A   B   I   G   P   U   D   D   I   F
5   9   3   11   2   15   5   17   14   8   18   18   13   10

Answer: W   E   T
16   10   3

Troubled Triangles

Grade Levels: 3 - 6

Number of Participants: any number

Materials Required: Spirit duplicating master

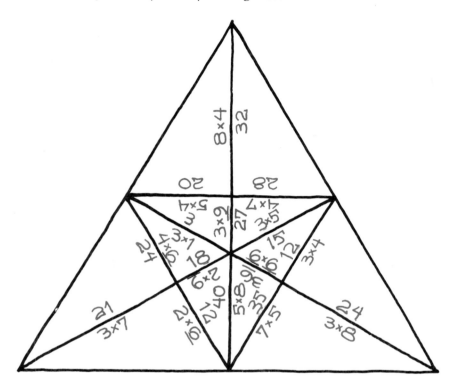

Directions:

1. Paste puzzle on posterboard and cut neatly along the dark lines.

2. Scramble the pieces.

3. Fit the triangles together so that all touching edges name the same number.

4. If all pieces form one large triangle, the answers are correct.

Circle the Globe

Grade Levels: 3 - 5

Number of Participants: any number

Materials Required: Chalk and chalkboard

Directions:

1. Divide the participants into teams of equal ability.

2. Draw a circle on the board for each team.

3. Place the numerals from 0 - 9 at random on the inside rim of each circle.

4. Have the first player from each team come to the board. (Try to have students of equal ability compete against each other in every round.)

5. Give a numeral from 0 - 5 for the players to place in the middle of the the circle.

6. The players will multiply the number in the center of the circle by the numbers on the inside rim of the circle and place their answers on the outside rim of the circle.

7. The first player to multiply all of the numbers by the center number receives 5 points for her or his team; the runner up receives 3 points, and the third place winner gets 1 point.

8. The team that has the most points at the end of the game is the winner.

Sum It Up

Grade Levels: 3 - 5

Number of Participants: any number

Materials Required: Chalk and chalkboard

Directions:

1. Divide the participants into teams of equal ability.

2. Each team lines up at the board.

3. At a given signal the first player on each team writes a multiplication problem with its product.

   NOTE: No product may be over 45.

4. As the first player finishes, he or she hands the chalk to the next player on the team who writes a different multiplication problem with its product on the board.

5. Players take turns, making sure no problem is duplicated.

6. The object is to see which teams can write as many different combinations on the board in a given time period.

7. When time is called, the products are added and the winning team would be the one with the largest sum.

NOTE: There can be no duplication of problems when team is computing its final score.

Pondered Products

Grade Levels: 3 - 5

Number of Participants: any number

Materials Required: Playing mat or spirit duplicating master, number cubes

X	1	2	3	4	5	6
1						
2						
3						
4						
5						
6						

Directions:

1. Each player prepares his or her own playing mat.

2. Each player places a multiplication sign in upper left hand corner and writes the numerals 1 - 6 in sequence horizontally and vertically on his or her own playing mat.

3. The participants roll a cube, and the player with the highest number goes first.

4. The first player rolls the number cubes and multiplies the two numbers shown.

5. If product is correct, the player puts the product in the appropriate square on his or her playing mat.

6. Players take turns.

7. The winner is the first player to fill in a row, column, or diagonal on his or her playing mat.

IDEA: 2 - 6 players may put numbers in random order instead of in sequence.

IDEA: If the entire group plays, then have pupils place numbers randomly. Teacher would call out the two numbers and pupils would place product where it is appropriate. Winner would be first player to get a row, column, or diagonal filled.

Five Straight

Grade Levels: 3 - 6

Number of Participants: 2

Materials Required: Gameboard, two different colored sets of markers (for example: white and blue chips, buttons or bottle caps)

2×9	3×8	4×5	3×3	5×8	2×4	3×7
6×4	5×4	2×8	4×4	5×3	4×7	5×7
8×5	5×5	6×3	4×6	2×2	6×7	2×7
8×3	2×3	3×4	5×2	3×2	5×6	4×8
9×4	4×9	6×6	4×3	6×2	3×5	5×9
8×4	3×9	6×5	7×3	6×8	2×5	6×1
4×2	8×2	7×5	2×6	7×2	7×4	3×6

Directions:

1. The first player chooses any square on the board and gives the factors and product.

2. If the player is correct, he or she places his or her marker over that space.

3. If the player is incorrect, he or she loses the turn.

4. The second player takes a turn.

5. The winner is the first player to cover 5 squares in a row, column, or diagonal.

IDEA: To play with more than two pupils, divide participants into two teams of equal ability. Play proceeds as above but members of the team take turns trying to place the team's markers on the playing board.

The Lone One

Grade Levels: 3 - 6

Number of Participants: 3 - 5

Materials Required: 20 cards with multiplication facts, 20 cards with products for multiplication facts, 1 card: The Lone One

Directions:

1. Shuffle the cards and deal all of the cards to the players.

2. The first player draws a card from the player on her or his left and then lays down any pair in his or her hand; e.g., 3 x 5 and 15 would be a pair.

3. Players take turns.

4. The winner is the first player to pair all of his or her cards.

5. The loser is the player left with the "Lone One".

NOTE: These cards (minus "The Lone One") could be used for the activity, Multi-Match.

IDEA: Instead of drawing a card from the player on his or her left, players could pass a card to the player on the right.

Checkerboard

Grade Levels: 4 - 6

Number of Participants: 2

Materials Required: Gameboard, 64 squares to cover gameboard (one side containing problem, other side containing product, divide the squares into 2 sets)

2×2	6×4	1×1	6×8	3×3	2×7	6×6	4×7
5×5	9×9	9×7	10×10	4×4	4×4	2×9	5×5
8×8	8×4	7×7	8×7	6×3	3×7	5×8	9×9
5×8	9×3	7×7	2×10	6×9	3×11	10×10	4×12
8×6	3×9	7×4	2×2	6×7	3×3	5×6	7×4
8×9	4×7	9×4	3×7	7×1	2×5	3×8	1×8
2×4	3×9	9×5	8×8	8×4	6×6	7×9	5×9
8×7	3×6	9×5	4×9	3×4	5×7	4×5	6×8

Directions:

1. Prepare gameboard (as shown above) and squares. It may be helpful to color every other square (like a checkerboard) before putting problems on gameboard. All problems facing in one direction would be in one color, all problems facing in the other direction would be in another color.

   NOTE: Instead of preparing gameboard an old checkerboard could be used with problems written directly on checkerboard.

2. Prepare squares to cover each problem. Products would be on one side and problem would be on the other side.

   NOTE: Shade the front and back of one set of cards to correspond to the shaded sections of the gameboard.

3. The two players sit across from each other and work only with the problems facing them.

4. At a given signal each player starts to cover each combination with the correct product.

5. The player who covers all of her or his combinations is the winner.

6. The players check the combinations by turning the squares over to see if the problem on the square matches the problem on the gameboard.

7. Players exchange sides and repeat above.

IDEA: Products could be written on board and small squares could contain the multiplication facts.

Point-a-Fact

Grade Levels: 3 - 5

Number of Participants: any number

Materials Required: Chalkboard, chalk and pointer for each team (ruler could be used)

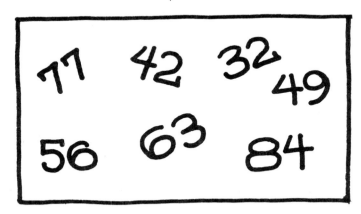

Directions:

1. Divide the participants into teams of equal ability.

2. The products of the multiplication facts in which the players need drill are placed on the board. The same products are used for each team.

3. The first player from each team comes to the board.

4. Each player is given a pointer.

5. The leader gives a problem such as 7 x 7 (the products of the problems given should be on the board).

6. The first player to point to the correct product gets a point for her or his team.

7. The pointer is passed to the next player on each team and the process is repeated.

8. The winning team is the team with the most points at the end of the game.

IDEA: Each player is given a piece of chalk and circles the correct product instead of pointing to it.

22

Multi-Match

Grade Levels: 4 - 6

Number of Participants: 1 - 3

Materials Required: 12 multiplication fact cards (without products)
12 answer cards

Directions:

1. Shuffle the cards and lay them in a 6 x 4 array, face down.

2. The first player turns over two cards (one at a time). If the cards match; e.g., multiplication fact goes with product, the player keeps both cards.

3. If the facts do not match, the player returns the two cards face down on the table.

4. The second player turns over two cards (one at a time) trying to locate a match. If the cards match, the player keeps them. If not, they too are returned to the table.

5. The play continues until the table has been cleared.

6. The winner is the player with the most cards at the end of the game.

Around the Room

Grade Levels: 4 - 6

Number of Participants: any number

Materials Required: Flash cards with multiplication problems

Directions:

1. The first two players are shown a multiplication card by the teacher.

2. The first player to give the product correctly travels to the next person's desk while the loser sits down.

3. The "traveler" and the next player compete to give correctly the product of the new fact card that the teacher holds up.

4. The winner proceeds to the next player and the loser sits down.

5. Play proceeds as described above.

6. The winner is the first person to travel around the room and reach her or his original seat.

Infield Out

Grade Levels: 4 - 6

Number of Participants: any number

Materials Required: Spirit duplicating master

Directions:

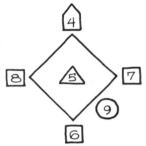

1. Given the infield as pictured, can you make a "put out" by naming the products obtained?

2. Each position is given a number. The diagram indicates play. The first one has been done for you.

3. Each player continues until an incorrect answer is given.

IDEA: Diagram could be put on board and game could be played orally by dividing group into teams of equal ability.

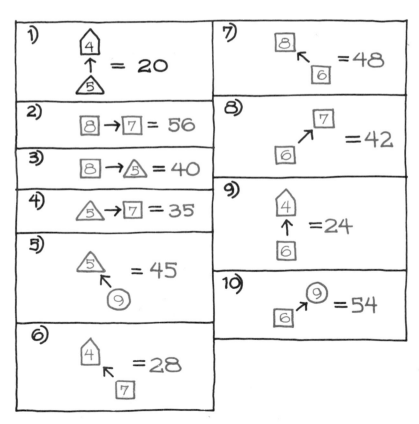

Solomon's Way

Grade Levels: 4 - 6

Number of Participants: any number

Materials Required: None

Directions:

1. Any single digit number greater than five can be multiplied by any other single digit number greater than five without knowing the tables beyond 5 x 5.

2. To illustrate, let's take the multiplication fact 8 x 7.

3. Write 8 and 7 as shown.

$$\boxed{\begin{matrix} 8 \\ 7 \end{matrix}}$$

4. Now subtract each number from 10 and write the difference on the side.

$$\boxed{\begin{matrix} 8 & \quad 2 \\ 7 & \quad 3 \end{matrix}} \qquad \begin{matrix} 10 - 8 = 2 \\ 10 - 7 = 3 \end{matrix}$$

5. The tens digit in the answer can be found by subtracting either number diagonally; e.g., 8 − 3 = 5 or 7 − 2 = 5. Therefore, the tens digit is 5.

$$\boxed{\begin{matrix} 8 & \diagdown & 2 \\ 7 & \diagup & 3 \end{matrix}}$$

6. The ones digit is found by multiplying the numbers on the right. The ones digit is 6 as 3 x 2 = 6. Thus 8 x 7 = 56.

Now you try some:

8 x 8	8 x 9	9 x 5	6 x 7
8 2	8 2	9 1	6 4
8 2	9 1	5 5	7 3

Cover Up

Grade Levels: 3 - 6

Number of Participants: 2 - 3

Materials Required: Playing mat, posterboard squares to fit playing mat (squares should have multiplication facts which correspond to product on mat plus some blank squares).

X	1	2	3	4	5	6	7	8	9	10
1	1	2	3	4	5	6	7	8	9	10
2	2	4	6	8	10	12	14	16	18	20
3	3	6	9	12	15	18	21	24	27	30
4	4	8	12	16	20	24	28	32	36	40
5	5	10	15	20	25	30	35	40	45	50
6	6	12	18	24	30	36	42	48	54	60
7	7	14	21	28	35	42	49	56	63	70
8	8	16	24	32	40	48	56	64	72	80
9	9	18	27	36	45	54	63	72	81	90
10	10	20	30	40	50	60	70	80	90	100

Directions:

1. Place all squares face down.

2. Each player draws one square to determine order of play. The player who draws the square with the highest value plays first.

3. Return all squares face down.

4. Each player now draws 10 squares.

5. The first player matches a square with a product on the playing mat and places that square on the appropriate square on the playing mat.

6. The next player must play a square that touches the first covered square horizontally or vertically (not diagonally). If the player has a blank square, he or she can play it.

7. If the player cannot play one of his or her squares, he or she draws a square from the pile. The player loses that turn if the square can not be played.

8. Players take turns, remembering that each square played must touch one of the squares already played.

9. The winner is the first player to play all of his or her squares.

IDEA: On an empty grid have players put numbers in random order in the first row and column; then they are to fill in correct products and proceed as above.

Multi-Deal

Grade Levels: 4 - 6

Number of Participants: 3 - 4

Materials Required: 40 cards (Numbered from 1 - 10, playing cards with face cards removed could be used.)

Directions:

1. Players draw from the deck to determine which player will be the dealer. High card becomes the dealer.

2. The cards are shuffled by the dealer and placed face down in front of him or her.

3. The dealer turns over two cards at a time and the first player who gives the correct product of these two numbers keeps the cards.

4. If there is a tie, each person gets one card. The dealer makes all decisions.

5. The winner is the player with the most cards at the end of the game.

6. The winner becomes the new dealer.

Given a 1 digit factor and a 2 - 4 digit factor with no renaming, the pupil will be able to find the product.

Multi-Cross Number

Grade Levels: 3 - 5

Number of Participants: any number

Materials Required: Spirit duplicating master

Directions:

Solve the cross number puzzle by placing the product in the correct space.

Across	Down
a) 2131 x 3 =	a) 324 x 2 =
d) 43 x 2 =	b) 3201 x 3 =
f) 11 x 4 =	c) 13 x 3 =
g) 23 x 3 =	d) 430 x 2 =
h) 21 x 3 =	e) 635 x 1 =
i) 3420 x 2 =	f) 23 x 2 =
j) 101 x 5 =	j) 10 x 5 =
k) 100 x 3 =	

Erase-It

Grade Levels: 3 - 5

Number of Participants: any number

Materials Required: Chalk and chalkboard

Directions:

1. Teacher will write on the board as many problems as there are pupils in the group.

2. Problems should be a 2 - 4 digit numeral multiplied by a 1 digit numeral with no renaming necessary.

3. Divide the group into three teams of equal ability.

4. First person on each team will go to the board and work any problem written on it.

5. Once the first player finds the correct answer, he or she erases the problem and gives the chalk to the next member of the team.

6. The winning team is the first team to have each of its members do one problem on the board.

NOTE: Team's members should watch the other teams multiplication to make sure no errors are made. If an incorrect product is written, then the person making the error must correct it.

Pass Thru

Grade Levels: 3 - 5

Number of Participants: any number

Materials Required: Spirit duplicating master

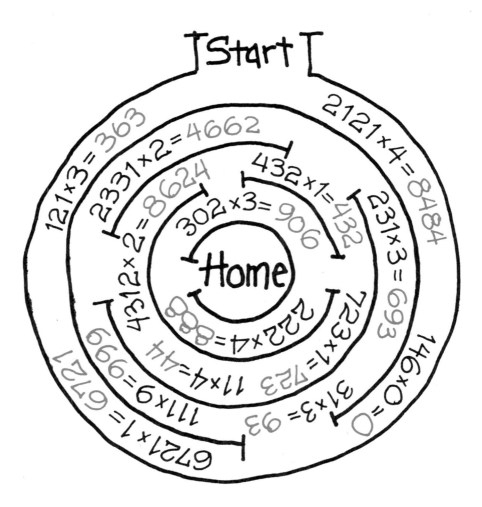

1. Get home by solving the problems blocking your way.

2. You can't pass without giving or writing a correct product.

3. There are several paths you can choose.

Given a 1 digit factor and a 2 - 4 digit factor with renaming, the pupil will be able to find the product.

Finish Line

Grade Levels: 3 - 6

Number of Participants: 2 - 4

Materials Required: Gameboard, markers, number cube (die)

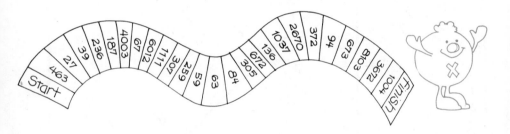

Directions:

1. Roll the number cube. The person rolling the highest number goes first.

2. The first player rolls the number cube and moves his or her marker the number of spaces shown on the cube.

3. The player then multiplies the number on the cube with the number on which the marker lands.

4. If the answer is correct, player remains on the square; if the answer is incorrect, the player must move her or his marker back to the starting square.

5. The player who reaches the finish line first wins.

NOTE: If you want the players to multiply by numbers 4 - 9, a die can be covered with masking tape and numbers can be written on it with a felt pen. If a calculator can be made available, a referee could use the calculator to check all work by the players.

Puzzle-Plication

Grade Levels: 4 - 6

Number of Participants: any number

Materials Required: Posterboard, pen and scissors

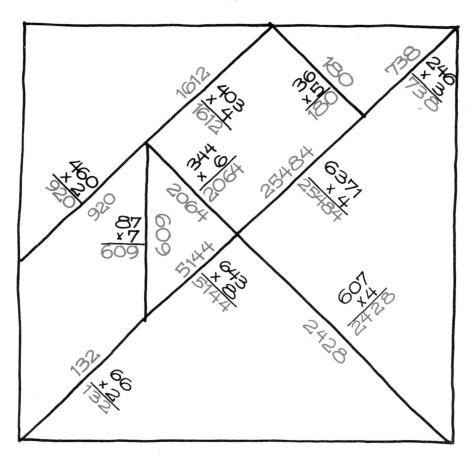

Directions:

1. Draw the puzzle shown above on the posterboard. Cut out the pieces carefully.

2. Mix the puzzle pieces.

3. The puzzle can be solved by answering each multiplication problem, then matching equal answers on the different pieces.

4. The seven pieces in this puzzle will form a square if they are put together correctly.

Multi-Magic

Grade Levels: 4 - 6

Number of Participants: any number

Materials Required: Chalk and chalkboard or paper and pencil

Directions:

A. Strange Fact

   1. Multiply 123,456,789 x 8.

   2. Add 9 to your answer.

   3. What did you get?   *987654321*

B. Favorite Number

   1. Pick out your favorite number from 1 - 9.

   2. Multiply 12,345,679 by your favorite number.

   3. Take that product and multiply it by 9.

   4. What did you get?

     Your favorite number again and again.

C. Magic Five

   1. A quick way to multiply by 5 is to multiply the other factor by 10 and then divide the answer by 2.

   2. For example, take 3,468 x 5.

     a. Think 5 is another name for 10/2.

     Hint: An easy way to multiply by 10 is to place 0 at the end of the number.

     b. Multiply 3,468 x 10 to get 34,680.

     c. Divide this number by 2.

$$\frac{34,680}{2} = 17,340$$

     d. Therefore:

$$3,468 \times 5 = \frac{34,680}{2} = 17,340$$

3. See if you can do the following problems in the same way.

$$2673 \times 5 = 13,365$$
$$6,486 \times 5 = 32,430$$
$$3,972 \times 5 = 19,860$$
$$673 \times 5 = 3,365$$
$$8,764 \times 5 = 43,820$$

Flip-It

Grade Levels: 3 - 6

Number of Participants: 2 - 5

Materials Required: 12 cards with the numbers 63, 24, 11, 221, 86, 127, 58, 77, 15, 46, 33, 148 on them, number cube (die) paper and pencil.

Directions:

1. Each player rolls number cube with the highest number going first.

2. Shuffle the cards and place them face down.

3. The first player throws the number cube and draws one card.

4. Multiply the number on the card by the number on the cube. Record the product.

5. Players take turns.

6. On the players second turn, the product obtained is added to the first product. On the third throw the product is added to the sum of the first two products, etc. Each player records his or her score.

7. The first player to reach a predetermined sum; e.g., 1,000 is the winner.

8. If all the cards are used, shuffle them again and place them face down.

IDEA: If a certain number appears in your sum; for example, 3, you lose all of your points and have to start again.

The Traveling Angles

Grade Levels: 4 - 6

Number of Participants: any number

Materials Required: Spirit duplicating master or chalk and chalkboard

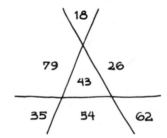

Directions:

1. Look at the numbers in the figure above. Each number is contained in a specific shape.

2. Match the shape in the above figure with the shape in the equations.

3. Write the appropriate number in the shapes of the equations.

4. Solve the equations.

$$35 \times 62 = 2{,}170 \qquad 43 \times 26 = 1{,}118$$

$$35 \times 18 = 630 \qquad 18 \times 43 = 774$$

$$62 \times 43 = 2{,}666 \qquad 18 \times 54 = 972$$

$$26 \times 18 = 468 \qquad 26 \times 79 = 2{,}054$$

$$35 \times 54 = 1{,}890$$

$$43 \times 54 = 2{,}322$$

The Puzzler

Grade Levels: 4 - 6

Number of Participants: any number

Materials Required: Spirit duplicating master

Directions:

1. Solve the multiplication problems.

2. Find the answer (product) in the letters listed above.

3. Whenever you see the number below a space, put the correct letter on the line in the space.

4. For example, the answer to #1 is 918. That number is found in the letter T. Therefore, wherever you see the number 1 put the letter T in the empty space.

M   U   L   T   I   P   L   I   C   A   T   I   O   N :
__ __ __ __ __ __ __ __ __ __ __ __ __ __
2   12 7   1   4   13 7   4   9   5   1   4   10 8

        L   A   Z   Y   M   A   N ' S
        __ __ __ __   __ __ __ __
        7   5   6   3   2   5   8   14

        A   D   D   I   T   I   O   N .
        __ __ __ __ __ __ __ __
        5   11 11 4   1   4   10 8

36

1. 27 x 34 = 918
2. 18 x 12 = *216*
3. 35 x 24 = *840*
4. 45 x 20 = *900*
5. 62 x 36 = *2,232*
6. 11 x 28 = *308*
7. 36 x 21 = *756*

8. 92 x 24 = *2,208*
9. 34 x 62 = *2,108*
10. 63 x 41 = *2,583*
11. 72 x 18 = *1,296*
12. 25 x 30 = *750*
13. 54 x 46 = *2,484*
14. 74 x 25 = *1,850*

One Digit Multiply

Grade Levels: 5 - 6

Number of Participants: any number

Materials Required: Paper and pencil or chalk and chalkboard

Directions:

1. If you like to multiply by one digit instead of two, here is how it can be done.

   a. 16 x 25 can be changed to 8 x 50. Take ½ of the first number and double the second factor and multiply. Multiply both problems and you will see that you get the same answer.

   b. 36 x 25 can be changed to 6 x 150. Take 1/6 of 36 and multiply the 25 by 6. Multiply these two problems and see if you don't get the same answer.

2. Try the following problems. Check them by multiplying the way you were taught.

   a. 18 x 55 = *990*

   b. 12 x 32 = *384*

   c. 14 x 67 = *938*

   d. 24 x 35 = *840*

   e. 25 x 62 = *1550*

Eye Openers

Grade Levels: 5 - 6

Number of Participants: any number

Materials Required: Chalk and chalkboard or paper and pencil

Directions:

I. The Half-double Method

    a. Any two 2 digit numerals can be multiplied by the following method:

       1. Take any two numbers; e.g., 45 x 34.

       2. Take ½ of the first number (drop all remainders). Double the second number.

22 x 68

       3. Continue to do so until you reach the number 1 in the first column.

11 x 136

5 x 272

2 x 544

1 x 1,088

       4. Next, cross out all the numbers in the second column that are opposite an even number in the first column.

Column I		Column II
45	x	34
22- - - - - -x- - - - -		68
11	x	136
5	x	272
- 2- - - - - -x- - - -		544
1	x	1088
		1530

       5. Add all the numbers in the second column that are opposite the odd number in the first column. (The ones not crossed out.)

       6. The answer to 45 x 34 is *1530*.

II. Instant Magic

a. To multiply two 2 digit numerals where the ones digits add up to 10 and the tens digits are the same; e.g., 34 x 36, do the following:

1. Take 34 x 36. Note that the ones digits add up to 10. (4 + 6 = 10). There are 3's in the tens column.

2. Multiply the ones column numbers together to get 24 (4 x 6 = 24). These will be the last two digits of your answer.

$$
\begin{array}{r}
34 \\
\times\,36 \\
\hline
24
\end{array}
\quad (4 \times 6 = 24)
$$

3. Increase one of the tens place factors by one and then multiply by the other tens place factor. For example,

$$
\begin{array}{r}
34 \\
\times\,36 \\
\hline
1{,}224
\end{array}
\quad (3 + 1) \times 3
$$

4. 34 x 36 = 1,224.

Increase one 3 to 4 and multiply the 4 by the other 3 (4 x 3 = 12). (This number will be the first half of your answer.)

b. Look at these problems and see if you can figure out how to get the answers.

1. **44 x 46 = 2,024**   4 x 6 = 24   4 x 5 = 20   *answer 2,024*

2. **38 x 32 = 1,216**   8 x 2 = 16   3 x 4 = 12   *answer 1,216*

3. **65 x 65 = 4,225**   5 x 5 = 25   6 x 7 = 42   *answer 4,225*

c. Now see if you can work these problems that same way:

1. **22 x 28** = *616*

2. **76 x 74** = *5,624*

3. **45 x 45** = *2,025*

4. **42 x 48** = *2,016*

5. **33 x 37** = *1,221*

III. Mental Products.

   a. To multiply any 2 digit number by another 2 digit number do the following:

   1. Take 36 x 43 for example:

   2. First multiply the ones (6 x 3 = 18). Put down the 8 and remember the 1.

$$
\begin{array}{r}
36 \\
\times\,43 \\
\hline
18
\end{array}
\qquad (3 \times 6 = 18)
$$

   3. Next cross multiply. Multiply the ones by the tens column. (3 x 3 = 9). Multiply the tens by the ones column. (4 x 6 = 24). Add the two products together (24 + 9 = 33) and also add the number you were remembering. (33 + 1 = 34).

$$
\begin{array}{cc}
36 \\
\times\,43 \\
\hline
48
\end{array}
\qquad
\begin{array}{r}
3 \times 3 = 9 \\
4 \times 6 = 24 \\
\hline
33 \\
+1 \\
\hline
34
\end{array}
$$

   4. Put down the 4 and remember the 3.

   5. Multiply the tens column by the tens column and add the number you were remembering. (3 x 4 = 12 + 3 = 15). Put down the 15.

$$
\begin{array}{r}
36 \\
\times\,43 \\
\hline
1,548
\end{array}
\qquad
\begin{array}{r}
3 \times 4 = 12 \\
+3 \\
\hline
15
\end{array}
$$

   b. Make up some 2 digit by 2 digit multiplication problems and try doing them this way. Check your answer by multiplying in the way you were taught.

Multiplication - 9

Given any two 2 digit factors, the pupil will be able to show that the order in which they are multiplied does not affect the product.

Petal Products

Grade Levels: 5 - 6

Number of Participants: any number

Materials Required: Spirit duplicating master

Directions:

1. Complete the multiplication flowers by multiplying the number in the center of the flower by the numbers in the petals around the center of the flower.

2. Put your answer in the outer row of petals.

3. Multiply the number in the petals around the center by the number in the center of the flower. Are the answers the same?

The Broken Field Runner

Grade Levels: 5 - 6

Number of Participants: any number

Materials Required: Spirit duplicating master

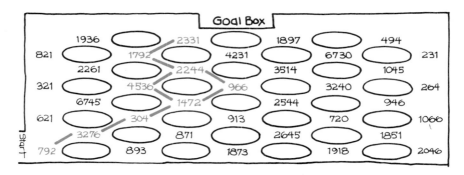

Directions:

1. Each passageway in this maze contains a number.

2. Fifteen of these numbers are the products to the numbers in the boxes below. The factors can be put in any order when you multiply.

3. Work any problem and, as you find the product in the maze, circle it.

4. Keep on working problems until you can draw a path from the entrance to the goal. Note the numbers circled form the path you may travel.

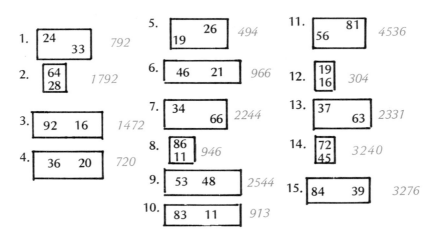

42

Cube Dictation

Grade Levels: 5 - 6

Number of Participants: 2 - 6

Materials Required: Four number cubes with digits 0 - 9 on them,
                     paper, pencil

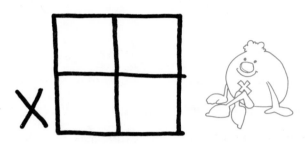

Directions:

1. Each player draws a box as shown on a sheet of paper.

2. Roll each of the number cubes one at a time and place the indicated
   number in one of the squares in the box. The object being to place
   the indicated number in the square so that the largest possible product is obtained when the squares are filled and the indicated multiplication is performed.

3. When all four squares are filled, perform the indicated multiplication.
   The player with the largest product wins one point. If two or more
   get the largest product, each wins a point.

4. The first player to get 10 points is the winner.

*Given any three or more 1 digit factors, the pupil will be able to show
that the way in which they are grouped does not affect their product.*

Zingo

Grade Levels: 3 - 6

Number of Participants: any number

Materials Required: Spirit duplicating master (could be put on
chalkboard)

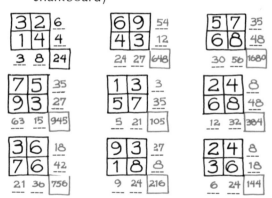

Directions:

1. Multiply the numbers horizontally, and put your answers in the
   blanks provided to the right of the squares.

2. Multiply the numbers vertically, and put your answers in the blanks
   provided underneath the squares.

3. Now multiply the two products that you got when you multiplied
   horizontally and put the new product in the square.

4. Next multiply the two products that you got when you multiplied
   vertically and you should have arrived at the same product that is in
   the square.

5. The first one is done for you.

6. See how many of the others you can do.

Three Cubes

Grade Levels: 4 - 6

Number of Participants: 3 - 4

Materials Required: 3 number cubes (dice), paper and pencil

Directions:

1. Each player rolls a cube with the highest roller being the leader.

2. The leader rolls the three cubes and the three players multiply the numbers shown on the cubes. (Players should use paper and pencil to do their computations.)

3. The first player who has the correct product for the three factors gets 3 points.

4. Each player who also gets the correct answer gets 2 points.

5. The leader checks the player's work, awards points, and keeps score.

6. At the end of the 5 rolls the player with the most points becomes the new leader.

NOTE: Students should notice that each player probably solved the problem by different groupings. If there are only 2 players, the role of the leader is eliminated.

IDEA: Instead of points, each player keeps track of products for each round he or she wins. At the conclusion of game each player adds his or her products. The winner is the player whose products add to the greatest sum.

The Web

Grade Levels: 4 - 6

Number of Participants: any number

Materials Required: Chalk and chalkboard or spirit duplicating master
and pencil

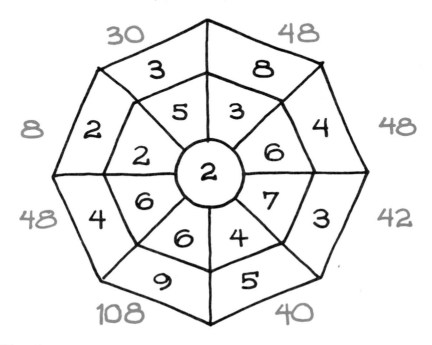

Directions:

1. Divide the participants into teams of equal ability.

2. Draw a figure similar to the one above for each team.

3. Fill in the web with numbers appropriate to the ability of the class.

4. At a given signal the first player from each team multiplies the
   number in the middle by the numbers going in a straight line from
   the center. This player places the answer on the outside edge of the
   web, and then solves it again from the outside to the middle to check
   the answer.

5. After completing one problem, the first player hands the chalk
   to the next member of the team and that player will do another
   problem in the web.

6. The players take turns until the web is completed.

> NOTE: Some students may do more than one problem if their team has less than 8 members on it.

7. When all of the problems in the web are completed, any student may change the answer of the problem she or he has worked.

8. Each team will get 1 point for each correct answer, but will subtract one point for each incorrect answer on the web.

9. The first team who completes the web correctly will get 5 points. Any other team that correctly completes the web will get 3 points.

10. The winning team is the team with the most points at the conclusion of the game.

> NOTE: Problems of this type could be given for seat work.

Given a statement with a missing factor where multiplication distributes over addition, the pupil will be able to find the missing factor.

Jingle Match

Grade Levels: 5 - 6

Number of Participants: any number

Materials Required: Spirit duplicating master

Directions:

1. Use the distributive property of multiplication over addition to simplify each expression in Set A.

2. Find the simplified expression in Set B.

   Example: (9 x 24) + (3 x 24) = (12 x 24)

3. Each time you find two boxes with the same value, transfer the word from the top box to the bottom box.

4. When you are finished, you will be able to answer a question.

Set A

(9×24)+(3×24) it	(3×17)+(4×17) Flies?	(3×27)+(4×27) kicked	(3×14)+(20×14) is	(25×73)+(20×73) one
(24×236)+(10×236) out!	(35×71)+(25×71) them	(5×10)+(3×10) Question:	(34×62)+(20×62) knock	(13×42)+(20×42) of
(3×14)+(8×14) you	(12×15)+(2×15) Why	(18×72)+(10×72) it	(12×10)+(12×10) to	(21×54)+(20×54) ever
(9×25)+(10×25) unsafe	(21×47)+(20×47) If	(13×25)+(13×25) shoo	(20×49)+(35×49) could	(25×62)+(10×62) you,

Set B

Question: 8×10	Why 14×15	is 23×14	it 12×24	unsafe 19×25
to 24×10	shoo 26×25	flies? 7×17	If 41×47	one 45×73
of 33×42	them 60×71	ever 41×54	kicked 7×27	you, 35×62
it 28×72	could 55×49	knock 54×62	you 8×14	out! 34×236

48

Distribute - Time

Grade Levels: 5 - 6

Number of Participants: any number

Materials Required: Paper and pencil

Directions:

1. Have students compute 46 x 5.

2. Next have students compute 46 x 20.

3. Have them add the sum of the two products.

4. Now have the students compute 46 x 25. They should have the same answer.

5. Demonstrate several problems this way.

6. Assign several problems with the option of using the distributive property. Assign some where it would be an advantage to break the problem down and others where there would be no advantage.

Examples: (37 x 23) + (63 x 23)   (8 x 12) + (8 x 8)

7. Now give the students 5 problems where there would be an advantage to break the problems down.

Examples: (13 x 45) + (7 x 45)   (16 x 37) + (4 x 37)

   a. Have students do the problems without using the distributive property and time them.

   b. Have students do the problems using the distributive property. Time them and see if any time is saved.

49

Maze Magic

Grade Levels: 5 - 6

Number of Participants: any number

Materials Required: Spirit duplicating master

Directions:

1. Find two expressions (one in the answer box and one in the 3 x 5 square) that have the same value.

2. If you find two, then put the letter that appears in the answer box in the circle in the other box.

3. Can you find your way out of the maze by connecting boxes in which there is a letter? You may travel vertically or horizontally, not diagonally.

A (2×12) + (3×12)	60
B (263×20)+(263×31)	13,413
C (47×3)+(47×20)	1,081
D (36×21)+(36×30)	1,836
E (124×3)+(124×3)	744
F (43×20)+(43×34)	2,322
G (16×17) + (16×10)	432
H (23×12)+ (23×10)	506
I (87×20) +(87×31)	4,437
J (18×21) + (18×21)	756
K (10×97)+(2×97)	1,164
L (16×15)+(16×20)	560
M (103×4)+(103×10)	1,442
N (124×3)+(124×10)	1,612
O (862×10)+(862×5)	12,930

(O) 862×15 12,930	(C) 47×23 1,081	(G) 16×27 432 — Finish
(B) 263×51 13,413	(O) 672×53 35,616	(O) 24×36 864
(N) 124×13 1,612	(F) 43×54 2,322	(D) 36×51 1,836
(O) 18×19 342	(O) 239×12 2,868	(J) 18×42 756
(A) 5×12 60 — Start	(H) 23×22 506	(L) 16×35 560

50

Multiplication - 12

Given a multiplication problem where one of the factors is 10, 100, or 1,000, the pupil will be able to find the product.

Riddle Me

Grade Levels: 4 - 6

Number of Participants: any number

Materials Required: Spirit duplicating master

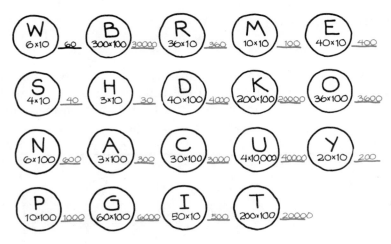

Directions:

1. Find the value of each letter by multiplying the two numbers together and placing the product on the line next to each circle. For example, the letter "W" has a value of 60.

2. Find 60 under one of the blanks and put the letter "W" above the 60.

3. Continue filling in the spaces with the letters of the correct products. Some letters will not be used.

4. When you are finished you will find the answer to the riddle.

When do boats become very affectionate?

$$\frac{W}{60} \quad \frac{H}{30} \quad \frac{E}{400} \quad \frac{N}{600} \qquad \frac{T}{20000} \quad \frac{H}{30} \quad \frac{E}{400} \quad \frac{Y}{200}$$

$$\frac{H}{30} \quad \frac{U}{40000} \quad \frac{G}{6000} \qquad \frac{T}{20000} \quad \frac{H}{30} \quad \frac{E}{400}$$

$$\frac{S}{40} \quad \frac{H}{30} \quad \frac{O}{3600} \quad \frac{R}{360} \quad \frac{E}{400} \; .$$

Rocket Trip

Grade Levels: 2 - 4

Number of Participants: 2 - 4

Materials Required: Gameboard, markers, 2 dice (3 - 8 on one and 10, 10, 100, 100, 1000, 1000 on the other) about 40 cards with the numbers 1 - 3 written on them (more 1 and 2 cards than 3 cards)

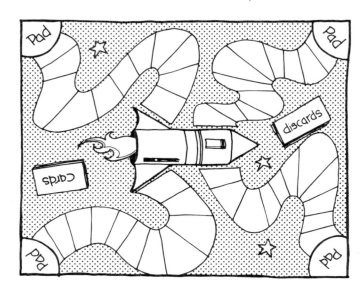

Directions:

1. Roll the 3 - 8 cube. The high roller is first.

2. Shuffle the cards and place them face down.

3. The first player rolls the two cubes and multiplies the 2 numbers showing.

4. If the player is correct, he or she draws a card and moves the marker that number of spaces on the playing board.

5. If the player is incorrect, he or she loses the turn.

6. Players take turns.

7. The winner is the player to reach the rocket ship first.

The Facts Man

Grade Levels: 1 - 3

Number of Participants: any number

Materials Required: Spirit duplicating master

X	10	100	1000	10	100	1000
3 X =	30	300	3000	30	300	3000
6	60	60	6000	600	600	600
9	90	90	9000	9	900	8000
7	70	9000	700	700	700	700
8	80	8000	8000	8000	800	800
6	60	60	6000	600	600	600
5	50	500	5000	50	500	5000

Directions:

1. The table shown above contains 16 errors.

2. Multiply the number in the top row by the number in the left column to get the answer given.

   Example: 10 x 3 = 30

3. As you find an error, rewrite the product and shade in the rectangle.

4. If you are correct, you will find a hidden message.

Palindrome Pals

Grade Levels: 4 - 6

Number of Participants: any number

Materials Required: Spirit duplicating master

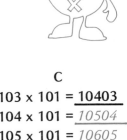

Directions:

1. Work at least two problems in each set.

2. Predict the next 2 or 3 answers but be careful.

3. Check your predictions.

A	B	C
101 x 101 = <u>10201</u>	102 x 101 = <u>10302</u>	103 x 101 = <u>10403</u>
101 x 202 = *20402*	102 x 202 = *20604*	104 x 101 = *10504*
101 x 303 = *30603*	102 x 303 = *30906*	105 x 101 = *10605*
101 x 404 = *40804*	102 x 404 = *41208*	106 x 101 = *10706*
101 x 505 = *51005*	102 x 505 = *51510*	107 x 101 = *10807*
101 x 808 = *81608*	102 x 808 = *82416*	108 x 101 = *10908*

D	E
111 x 111 = <u>12321</u>	121 x 121 = <u>14641</u>
111 x 222 = *24642*	121 x 131 = *15851*
111 x 333 = *36963*	121 x 141 = *17061*
111 x 444 = *49284*	121 x 151 = *18271*
111 x 555 = *61605*	121 x 161 = *19481*

Prove It

Grade Levels: 4 - 6

Number of Participants: any number

Materials Required: None

Directions:

1. A quick way to check multiplication problems is to "add the digits".
   To illustrate use the following problem.

$$\begin{array}{r} 673 \\ \times\ 236 \\ \hline 158,828 \end{array}$$

   a. Add the digits of the multiplicand $(6 + 7 + 3 = 16)$. Keep on adding
      the digits until only one number remains $(1 + 6 = 7)$. The digit to
      remember for the multiplicand is 7.

   b. Next add the digits of the multiplier. Keep on adding the answer
      until just one digit remains, $(2 + 3 + 6 = 11)$. Since 11 is still two
      digits add these numbers together, $(1 + 1 = 2)$. The digit to remem-
      ber for the multiplier is 2.

   c. Since there is a multiplication problem, you multiply the digit you
      got in the multiplicand by the digit you got in the multiplier,
      $(7 \times 2 = 14)$. This is the two digit number so you keep on adding
      the digits together until you get one digit, $(1 + 4 = 5)$. The digit to
      remember is 5.

   d. Now add the digits in the product. Keep on adding the digits of
      the sum until only one digit remains. If the answer is correct, it
      should be 5. $1 + 5 + 8 + 8 + 2 + 8 = 32$. $3 + 2 = 5$.

$$673 \longrightarrow 6 + 7 + 3 = 16 \longrightarrow 1 + 6 = 7$$
$$\times\ 236 \longrightarrow 2 + 3 + 6 = 11 \longrightarrow 1 + 1 = 2$$
$$\overline{} \qquad\qquad\qquad\qquad\qquad 14 \longrightarrow 1 + 4 = 5$$
$$158,828 \longrightarrow 1 + 5 + 8 + 8 + 2 + 8 = 32 \longrightarrow 3 + 2 = 5$$

55

2. Check the following problems to see if they are correct. One answer ;
is wrong, can you find it? *(Item "e")*

	a.	674	b.	234		c.	694
		x 236		x 361			x 324
		159,064		84,474			224,856

	d.	573		e.	704
		x 235			x 203
		134,655			142,902

I'm Done

Grade Levels: 5 - 6

Number of Participants: any number

Materials Required: Spirit duplicating master

785,204	80,476	609,245	284,700
114,648	325,306	504,145	325,476
67,003	82,004	162,866	138,306
154,507	214,245	174,870	158,296

Directions:

1. Work the problems.

2. Find the answers in the box and circle them.

3. Call out "I'm Done" when you have 4 circled answers in a line either
horizontally, vertically, or diagonally.

4. Continue playing until someone fills the sheet.

1. 236 x 341 = *80,476*   7. 523 x 622 = *325,306*
2. 673 x 242 = *162,866*   8. 376 x 421 = *158,296*
3. 876 x 325 = *284,700*   9. 562 x 204 = *114,648*
4. 247 x 332 = *82,004*   10. 621 x 345 = *214,245*
5. 623 x 222 = *138,306*   11. 967 x 812 = *785,204*
6. 435 x 402 = *174,870*   12. 845 x 721 = *609,245*

56

Given a composite number, the pupil will be able to identify the appropriate factors.

Factor Fun

Grade Levels: **5** - 6

Number of Participants: any number

Materials Required: Spirit duplicating master

Directions:

1. Cross out any letter which does not have under it a factor of the number in the circle.

2. Read the riddle and answer.

              ⑯

D Ⓦ O Ⓗ O C Ⓐ Ⓣ E D

3   2   5   4   7   3   4   8   6   9

  ㉔   ㉗

G Ⓟ R O Ⓔ Ⓣ W   Ⓒ O Ⓐ R Ⓝ S E

5   2   7   9   3   8 10   2   4   9   6   3   2   5

            ㉚

F Ⓨ E Ⓞ Ⓤ E D

7   5   8   6   2   4   9

  ㊱   ⑭

T Ⓢ H Ⓣ E Ⓐ Ⓝ Ⓓ   F Ⓞ U Ⓝ D ?

5   2   7   3   8   4   6 12   3   2   5   7   4

   Answer:

                ㉜

F A Ⓒ T Ⓐ Ⓡ Ⓟ A G Ⓔ Ⓣ

7   5   2 11   3   6   8 13 10   9 24

Riddle: *What pet can you stand on?*

Answer: *Carpet*

Prime Composites

Number of Participants: 3 - 6

Materials Required: 20 cards with composite numbers written in one felt pen color, 20 cards with prime factors of the given composite numbers written in a different felt pen color.

Directions:

1. Shuffle the prime number cards and place them face down on the table.

2. Shuffle the composite cards and deal the same number of cards to each player.

3. The first prime number card is turned over. If the prime number card is a factor of any composite card in a player's hand, she or he may discard that card.

4. Each successive prime number card is turned over and players try to play their composite cards on it.

5. The winner is the first player to play all the cards in his or her hand.

Ringo

Grade Levels: 4 - 6

Number of Participants: any number

Materials Required: Spirit duplicating master and pencil

Directions:

1. Have each participant make an array of 6 single digit numbers by 6 single digit numbers.

2. Leader calls out a composite number under eighty.

   NOTE: Keep track of numbers called to verify winner.

3. Students will ring the numerals that are next to each other either vertically or horizontally if these numbers are composed of the two factors that will equal the number called.

4. The first player to get 6 numbers ringed either vertically, horizontally, or diagonally will call out "Ringo".

5. The numbers ringed are checked with the leader's chart.

6. If correct, that person is declared the winner.

   Example:
   The leader can call "30" and the player would ring the 6 and 5 in the third row. The next call might be "16" and the player would ring the 8 and 2 in the third row. Play continues.

NOTE: Have the leader prepare a list of composite numbers before the game begins.

Star Bright

Grade Levels: 4 - 6

Number of Participants: any number

Materials Required: Spirit duplicating master

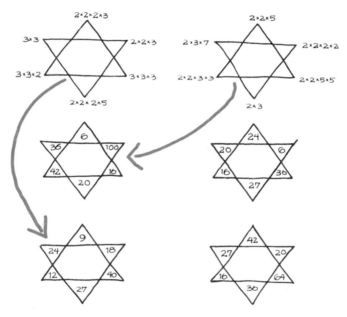

Directions:

1. Multiply each problem and place the product in the appropriate triangle.

2. Match the factored star with the star containing the correct composites.

60

Snowball

Grade Levels: 4 - 6

Number of Participants: any number

Materials Required: Spirit duplicating master

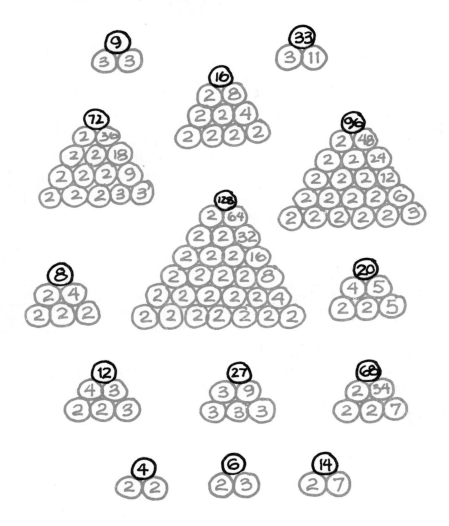

Directions:

1. Find the number in the top snowball. It is a composite number.

2. Factor the number in the top snowball. Write the answers in the empty snowballs. The numbers in the bottom row of snowballs must be the prime factors of the top number.

Primed

Grade Levels: 4 - 6

Number of Participants: 1 - 3

Materials Required: 14 prime number cards written in one felt pen color marked as follows: 2, 2, 2, 2, 2, 3, 3, 3, 5, 5, 7, 7, 11, 13, and 35.
Composite number cards written in another felt pen color as follows: 4, 6, 8, 9, 10, 12, 14, 15, 16, 18, 20, 21, 22, 24, 25, 27, 28, 30, 32, 33, 35, 36, 40, 42, 44, 45, 48, 49, 50, 52, 54, 55, 56, 60, 65, 70.

Directions:

1. Each participant turns over one card with the person drawing the highest composite card going first. Put the cards back in the deck.

2. Let the players refer to the prime number cards during the game.

| 2 | 13 | 11 | 2 | 3 | 5 | 2 |
| 2 | 7 | 7 | 2 | 3 | 5 | 3 |

3. Shuffle the composite cards and place them face down.

4. The first player turns over the first composite card and if he or she can (in a given time) name the prime factors of the card he or she keeps it. The player must say all of the products in the multiplication.

5. If he or she cannot give the prime factors of the composite number, the card is returned to the bottom of the deck. For example 24:
2 x 2 x 2 x 3 or 2 x 2 equals 4 x 2 equals 8 x 3 = 24.

6. Each successive player turns over the next card and proceeds as above.

7. The winner is the player with the most cards at the conclusion of the game.

IDEA: A solitaire game

1. Shuffle all cards together.

2. Lay out a 4 x 6 array.

3. The player may pick up any composite card as he or she locates all of its prime factors.

4. After all possible composite cards are picked up with prime numbers available, the player fills in empty spaces with cards from deck and proceeds as before.

Given a pair of numbers, the pupil will be able to name the least common multiple.

All the Way

Grade Levels: 5 - 6

Number of Participants: 2 - 5

Materials Required: Gameboard, markers, 2 number cubes (one with 2, 3, 4, 5, 6, 7, and one with 4, 5, 6, 7, 8, 9)

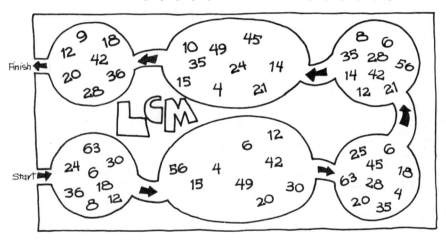

Directions:

1. Each participant rolls the number cubes and adds the two numbers showing. Player with the highest sum goes first.

2. The first player rolls the two number cubes and determines the LCM. If this number appears in the first circle, she or he places her or his marker in it.

3. If the number does not appear in the first circle, she or he does not move and passes the cubes to the next player.

4. Each player takes a turn.

5. The winner is the first player to travel around the gameboard stopping in each circle along the way.

IDEA: Change gameboard and numbers on the number cube.

IDEA: Play with 3 number cubes with students determining the LCM of the three numbers showing. Gameboard would have to be changed to accommodate new LCM's.

Multiple Relay

Grade Levels: 5 - 6

Number of Participants: any number

Materials Required: None

Directions:

1. Divide the group into teams of equal ability.

2. The first player from each team goes to the board.

3. The leader gives two or three numbers; e.g., 2, 3, 4.

4. The first player to write the correct least common multiple (12) is awarded 1 point for his or her team.

5. Each successive player from each team proceeds as in directions 2 - 4.

6. The team with the most points at the end of the game is the winner.

Multiple Deal

Grade Levels: 5 - 6

Number of Participants: 2 - 5

Materials Required: 40 cards (1 - 9 written on them) or playing deck with face cards removed.

Directions:

1. Shuffle the cards and place them face down.

2. If there are more than 2 players, one player will act as the dealer.

3. The dealer turns over the first two cards and the first player who gives the correct LCM of the 2 numbers keeps both of the cards.

4. If there is a tie, then each player keeps one card. The dealer will make all decisions.

5. At the conclusion of the game the player with the most cards becomes the new dealer.

IDEA: Instead of two cards being turned over, the three cards are shown to the players. The LCM of the three numbers is computed.

Given a multiplication problem in which both factors contain multiples of 10, the pupil will be able to find the correct product.

Zero into Home

Grade Levels: 4 - 5

Number of Participants: any number

Materials Required: None

Directions:

1. One player is selected to start the game with all other players seated in a circle. That player stands behind a player in the circle and the teacher gives a multiplication problem in which both factors contain multiples of 10.

2. The first player who answers the problem correctly moves on to the next player. The other player sits or remains in the chair.

3. Play continues as in directions 1 - 2.

4. The first player to return to his or her original seat is the winner.

Boxed in Factors

Grade Levels: 4 - 6

Number of Participants: any number

Materials Required: Spirit duplicating master and pencil

1800	60 × 30	30×600	3×600	90×20	900× 20
2400	30×800	20 × 1200	4×600	40×60	30×80
3600	90×40	300× 120	400×90	60×60	600×60
4000	2000×2	20×20	10×400	5×80	200×20
1200	30×40	300×40	30×600	20×60	10× 20

Directions:

1. Each row contains five problems. Work each problem and place the product under each rectangle.

2. Circle all of the problems whose product is the same as the number in the rectangle at the beginning of each row.

Puzzle Pieces

Grade Levels: 5 - 6

Number of Participants: any number

Materials Required: Tagboard, picture

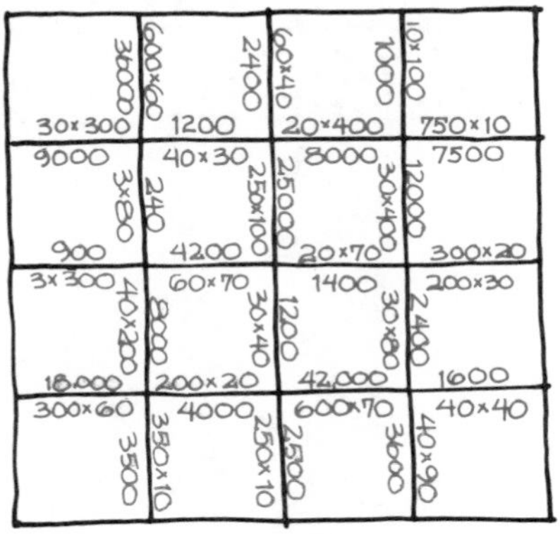

Directions:

1. Paste the puzzle on tagboard and the picture on other side. Cut it out carefully.

2. Mix the puzzle pieces.

3. Put the puzzle together by making sure that the edges that touch name the same number.

4. Turn the puzzle over and check to see that the picture is correct.

IDEA: You can make the puzzle self-checking by putting a letter in each square. For example,

Did you get it right?

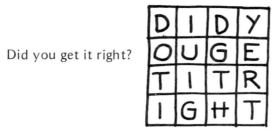

Then directions 1 through 3 would be the same and 4 would read — —

4. If your puzzle is correct, you will be able to read a message written especially for you.

Or letters could be placed in each box and students could check by comparing with the Answer Key. For example,

Answer Key

68

Given a multiplication problem involving any 1 or 2 digit factors, the pupil will be able to find the correct product.

Multi-Target

Grade Levels: 5 - 6

Number of Participants: any number

Materials Required: 10 cards (marked 0 - 9), pencil and paper

Directions:

1. Players draw the following grid on their papers.

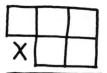

2. Cards are shuffled and placed face down.

3. The first card is turned over and the players place this number any where on their grid. The object is to get the largest product by doing the indicated multiplication when the squares are filled.

4. The second card is turned over and the players place this number on their grid.

5. Play proceeds until 5 cards have been turned over.

6. The players now multiply and the winner is the player with the largest product.

IDEA: Grid can be changed to:

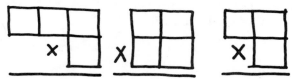

IDEA: Players can play with the same rules as above but try to obtain the smallest product.

IDEA: Players can play with the same rules as above but try to obtain a specific number.

Touchdown

Grade Levels: 4 - 6

Number of Participants: any number

Materials Required: Gameboard, chalk, chalkboard, or paper and pencil, and three stacks of cards

Red Cards: basic fact on one side, answer on other side.

Blue Cards: multiplication problem with no carrying involved on one side, answer on other side.

White Cards: multiplication problem with carrying involved on one side, answer on other side.

Directions:

1. Draw a replica of a football field on the board.

2. Divide the players into two teams of equal ability.

3. Designate goal posts for each team.

4. The first player on each team will compete against the first player on the other team. The teacher will start by giving the two players a multiplication fact. The first player who gives the correct answer will advance the ball 5 yards toward his or her goal post. (Ball starts on the 50 yard line.)

5. The second player from each team will now compete. The team carrying the ball will decide what type of multiplication problem will be given.

   a. Multiplication fact - advance 5 yards.

   b. Multiplication problem with no carrying - 10 yards.

   c. Multiplication problem with carrying - 15 yards.

6. If the team member carrying the ball gives a correct answer, he or she advances the ball the proper yardage.

7. If the team member carrying the ball gives an incorrect answer, he or she gives the ball to the opposite team.

8. If the team member not carrying the ball gives the correct answer first, then he or she receives the ball and has his or her turn.

9. If the team member not carrying the ball gives an incorrect answer, then the team carrying the ball has the option of not answering the problem and advancing 5 yards, or answering the problem already given and advancing (if correct) the yardage designated by the type of problem.

10. Paper and pencil may be used or work may be done on chalkboard.

11. Six points is given for every touchdown and the ball is returned to the 50 yard line.

12. The team with the most points at the end of the game is the winner.

NOTE: It may be wise to designate a certain time limit for the playing of this game.

Multiply Home

Grade Levels: 4 - 6

Number of Participants: 4 - 6

Materials Required: Gameboard, markers, 3 groups of cards, and one pair of dice.

Red Cards: contain the basic multiplication facts - problem on one side and problem and answer on the other side.

Blue Cards: contain multiplication problems that do not involve carrying - problem on one side and problem and answer on the other side.

White Cards: contain multiplication problems that involve carrying - problem on one side and problem and answer on the other side.

Red	Safe	White		White	Safe	Red			Red	Safe	Blue
Advance 2 spaces		Go Back 2 spaces		Red		Blue			Go back 1 Space		Red
Blue		Blue		Blue		Go Back 2 Spaces			Blue		Blue
White		Red		Advance 1 Space		White			White		White
White		Blue		Blue		Red			Blue		Red
Red		White		White		Blue			Red		White
Start		Red	Blue	Red		White	Blue	Advance 1 space			Home

Directions:

1. Each participant will roll dice with highest number going first.

2. The first player rolls the dice and advances that number of spaces on the gameboard.

3. If he or she lands on a red square, the player to his or her left draws a red card and gives him or her the problem that is written on it.

4. If the player answers the problem correctly, he or she moves forward one additional space. If he or she answers incorrectly, he or she must return to his or her original space.

5. If the player lands on a blue space, the player to his or her left draws a blue card and gives him or her the problem written on it.

6. If the player answers a blue card correctly, he or she moves forward an additional 2 spaces. If not he or she returns to his or her original space.

7. If the player lands on a white square, the player to his or her left draws a white card and gives him or her the problem written on it.

8. If the player answers a white card correctly, he or she moves forward an additional 3 spaces, if not he or she returns to his or her original space.

9. The first player to reach home is the winner.

Division Objectives

Objective	Pages

Fast Count

Grade Levels: 1 - 3

Number of Participants: any number

Materials Required: Counters (bottle caps, straws, buttons, etc.)

Directions:

1. Divide the group into teams of equal size.

2. Give each team the same number of counters.

3. Instructions: "If you give 5 people the same amount of counters, how many counters would each person receive?"

4. The first team which can pass them out and correctly name the number of counters each person received gets a point.

5. Repeat directions 2 and 3 varying the instructions and the amount of counters distributed to the teams.

6. The first team with 5 points wins.

NOTE: Both teams always receive the same number of counters.

Divisimo

Grade Levels: 1 - 3

Number of Participants: 2 - 4

Materials Required: Thirty dominoes constructed from poster board with 1 to 10 dots on each half of the card.

Directions:

1. Lay all the dominoes face down on the table.

2. In order to determine who plays first, each player draws one domino. The player who draws the highest domino total plays first.

3. Each player then draws five more dominoes, taking care not to reveal them to the other players at this time.

4. The first player turns over one of his or her dominoes.

5. The player to her or his left has a play if either end of one of the dominoes is a factor of either number.

   For example: If the first domino laid down shows 9 and 5, the second player must lay down a domino showing a number which divides evenly into 9 or 5.

6. If the second player cannot do this, he or she must draw one domino from the pile. If this domino does not divide evenly into either number, the player loses the turn and the next player tries.

7. Players take turns.

8. Play continues until one player who has used all of his or her dominoes is the winner.

Quarterback

Grade Levels: 1 - 3

Number of Participants: 3 or 5

Materials Required: Counters, cupcake tins or plastic rings from six packs

Directions:

1. One player acts as the quarterback.

2. Distribute to each player 10 counters and a cupcake tin or set of plastic rings.

3. Quarterback calls a number; such as, "6".

4. Each player places that number of counters in front of him or her.

5. The quarterback then calls a number smaller than the first; such as, "2".

6. The players divide the counters evenly into the number of cups indicated by the second call, (six counters divided evenly into two cups).

7. The players then call out how many counters are in each cup. (If the numbers 6 then 4 were called, the correct response would be, "One with 2 left over.")

8. The first successful player becomes the quarterback.

Come Home

Grade Levels: 3 - 5

Number of Participants: any number

Materials Required: Number line, picture of rabbits, and chart.

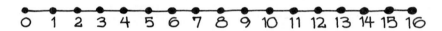

O 1 2 3 4 5 6 7 8 9 10 11 12 13 14 15 16

Start	No. of Jumps	Baby
12	4	3
15	3	5
16	2	8
16	4	4
14	2	7
6	1	6

Directions:

1. Mother Rabbit had so many children she couldn't remember their names. So she taught each baby rabbit to jump a certain size jump.

2. Rabbit 1 jumped only 1 space at a time. Rabbit 2 jumped only 2 spaces at a time. Rabbit 3 jumped only 3 spaces at a time and so on.

3. When mother called the babies for lunch she could tell which baby it was by counting the number of jumps it took the baby rabbit to get home, and then dividing the number of jumps into the distance the baby rabbit had to travel.

4. Can you help Mother Rabbit tell her children apart?

5. Fill in the chart for Mother Rabbit by writing the names of the babies on the lines at the right side of the chart.

"Egg" Citing Weights

Grade Levels: 3 - 4

Number of Participants: 1 - 2

Materials Required: Egg cartons, string, heavy washers, spirit duplicating master, and chart

No. of sections	No. of Washers	No. in each section	No. leftover
3	11		
	15		
	12		
	18		
4	12		
	8		
	16		
	18		
5	15		
	10		
	13		
	12		
2	4		
	18		
	6		
	12		
	14		
6	12		
	6		
	18		
	17		
	14		

Directions:

1. Cut egg cartons so that each section has at least one open side and is connected as shown:

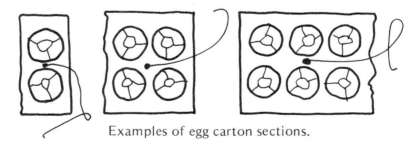

Examples of egg carton sections.

2. Attach a string to the center of each section.

3. Lift the carton by the string to make sure the carton is balanced.

4. Distribute a given number of washers to each student.

5. The student divides the number of washers by 2, 3, 4, etc., by placing the washers into each section of a specific egg carton as directed by the chart.

6. The student can check the answer by lifting the carton by the string in order to see if the egg carton balances.

Corralling the X's

Grade Levels: 2 - 4

Number of Participants: entire group

Materials Required: Number cube, chalkboard and chalk

Directions:

1. A tic-tac-toe grid is drawn on the chalkboard.

2. The group is divided into two teams of equal ability.

3. Two separate groups of "x's" are drawn on a separate part of the chalkboard.

4. The first player from each team goes to the chalkboard and stands in front of a group of x's.

5. The teacher rolls a number cube and calls out the number which appears.

6. The first player from each team groups the x's into groups the size of the number the teacher calls.

7. The first player to write the correct response on the board marks an "X" or "O" in a square.

8. The second player from each team goes to the board and directions 5 - 7 are repeated.

9. The winning team is determined by a tic-tac-toe win.

Given a division combination where the product is 45 or less, the pupil will be able to find the quotient.

Cut-Up Capers

Grade Levels: 3 - 5

Number of Participants: 1 - 2

Materials Required: Picture (approximately the same size as the puzzle), 3 pieces of cardboard and puzzle

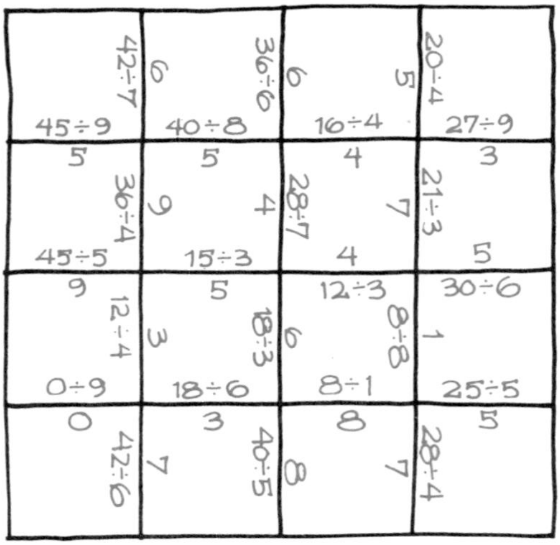

Directions:

1. Paste the picture on one side of the cardboard and the puzzle as shown above on the reverse side.

2. Cut the puzzle apart.

3. Mix up the pieces and have the student (using the number side) put the puzzle together on the cardboard base.

4. To check the puzzle, place another piece of cardboard over the puzzle then carefully turn it over. When the top cardboard is removed the complete picture will appear.

Travelin' Triangle

Grade Levels: 3 - 5

Number of Participants: any number

Materials Required: Spirit duplicating master

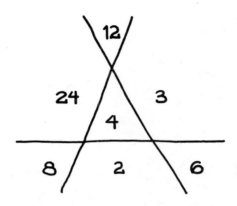

Directions:

1. Look at the numbers in the above figure. Each number is contained in a specific shape.

2. Match the shapes in the figure with the shapes in the equations.

3. Write the appropriate numbers in the shapes of the equations.

4. Solve the equations.

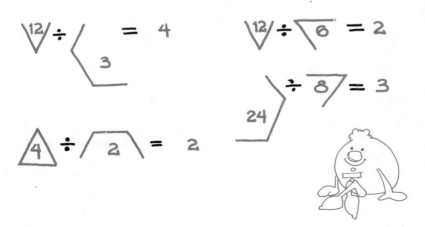

Waggin' Wheels

Grade Levels: 3 - 4

Number of Participants: any number

Materials Required: Spirit duplicating master

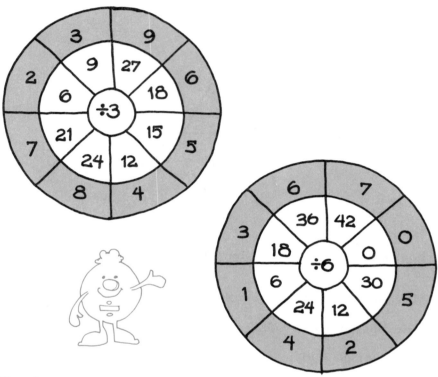

Directions:

1. Divide the center number into the number in the spokes.

2. Place the answer on the rim.

IDEA: Instead of using a spirit duplicating master divide group into teams of equal ability. Put "Waggin' Wheel" problems on the board (same wheel for each team) and have a race to see which team can complete the wheel first.

Password

Grade Levels: 4 - 5

Number of Participants: any number

Materials Required: Spirit duplicating master

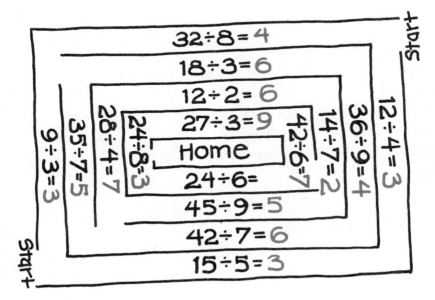

Directions:

1. Get home by answering the problems blocking your path.

2. You can't pass without giving a correct answer.

   (HINT: There are several ways to go.)

Given a multiple of 5 or 10 and a number of equal sets, the pupil will be able to find the number of members in each set.

Pyramid

Grade Levels: 3 - 4

Number of Participants: any number

Materials Required: Spirit duplicating master

Directions:

1. Climb the mountain to decode the message.

2. Substitute the correct letter for each numeral under the step.

a = 35 ÷ 5	i = 40 ÷ 4	r = 110 ÷ 10
c = 25 ÷ 5	l = 65 ÷ 5	s = 20 ÷ 5
d = 45 ÷ 5	m = 80 ÷ 10	t = 10 ÷ 10
e = 20 ÷ 10	n = 60 ÷ 5	u = 0 ÷ 5
f = 15 ÷ 5	o = 60 ÷ 10	v = 150 ÷ 10
h = 70 ÷ 5	p = 100 ÷ 5	

Spend Thrift

Grade Levels: 2 - 3

Number of Participants: any number

Materials Required: Spirit duplicating master

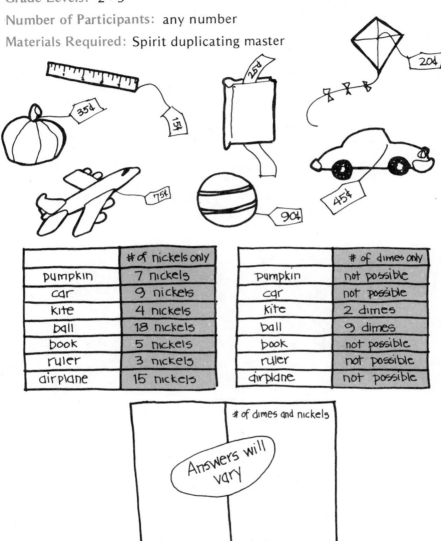

	# of nickels only
pumpkin	7 nickels
car	9 nickels
kite	4 nickels
ball	18 nickels
book	5 nickels
ruler	3 nickels
airplane	15 nickels

	# of dimes only
pumpkin	not possible
car	not possible
kite	2 dimes
ball	9 dimes
book	not possible
ruler	not possible
airplane	not possible

# of dimes and nickels

Answers will vary

Directions:

1. Barbara has only nickels and dimes to buy these things.

2. Fill in the charts to show how many dimes and how many nickels Barbara would need to buy each of these things. For Chart A and B you must use only nickels or only dimes.

NOTE: Children could use play money to help them solve problems.

Divide-A-Square

Grade Levels: 3 - 4

Number of Participants: any number

Materials Required: Cardboard and puzzle

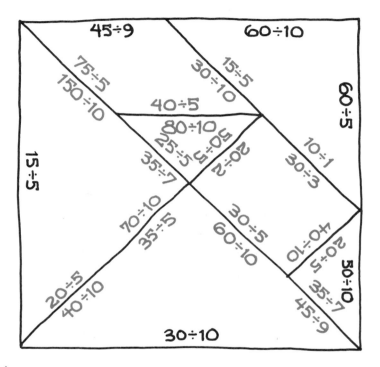

Directions:

1. Paste the puzzle on cardboard and trim the edges carefully.

2. Cut the puzzle along the lines.

3. Mix the puzzle pieces.

4. The puzzle can be solved by working each division problem and putting the pieces together whose sides contain problems that have equal answers.

5. The seven pieces in this puzzle will form a square if they are put together correctly.

Given any division problem based on a division fact; e.g., $9\overline{)72}$*, the pupil will be able to find the quotient.*

Divi-Disc

Grade Levels: 4 - 6

Number of Participants: 2 - 6

Materials Required: Hundreds board or gameboard, 100 key tags or playing discs with all products on them

X	1	2	3	4	5	6	7	8	9
1									
2									
3									
4									
5									
6									
7									
8									
9									

Directions:

1. One pupil with a multiplication chart is chosen to act as checker.

2. Tags or discs are placed face down on a table and mixed up.

3. Each player chooses 10 discs.

4. The first player turns a disc over and names two factors which produce that product.

5. This player then places the disc on the appropriate square on the board.

6. If a mistake is made, the player loses that turn and the player who discovers the mistake plays a disc.

7. If a zero or a multiple of 5 is drawn, a second disc is played.

8. The first player to use all 10 discs is the winner.

Divi-Trek

Grade Levels: 4 - 6

Number of Participants: 2 - 4

Materials Required: Gameboard, 4 markers, die, miniature gameboard with answers for reference if a dispute arises.

42÷6	49÷7	54÷6	48÷6	36÷4	10÷1	16÷8	24÷3
56÷7							27÷3
64÷8		81÷9	63÷9	21÷7	40÷5		30÷5
72÷8		18÷3			45÷5		32÷4
18÷6		15÷3			63÷7		48÷8
6÷2		0÷5			end		42÷7
5÷5		8÷1					56÷8
Start		9÷3	30÷6	36÷6	40÷8	45÷9	54÷9

Directions:

1. Each player rolls the die. The player rolling the highest number goes first.

2. The first player rolls the die to determine the number of spaces to advance.

3. The player must answer the problem correctly in each space passed as the marker is moved forward. If a mistake is made, the player must return to the space where he or she started the turn.

4. Players take turns.

5. The first player to reach the end space wins.

Cross-Number Puzzles in Division

Grade Levels: 4 - 6

Number of Participants: any number

Materials Required: Spirit duplicating master

16	2	8
8	2	4
2	1	2

30	6	5
5	1	5
6	6	1

18	6	3
3	1	3
6	6	1

12	4	3
6	2	3
2	2	1

40	8	5
2	2	1
20	4	5

27	9	3
9	3	3
3	3	1

48	8	6
6	2	3
8	4	2

42	7	6
2	1	2
21	7	3

54	6	9
9	3	3
6	2	3

Directions:

1. Divide both horizontally and vertically as if the division symbol were between each pair of numbers.

2. Divide the answers you get along the side. Divide the answers you get along the bottom row. The quotient for the two problems should be the same if your answers are correct.

Given any 4 digit numeral and a 1 digit divisor, the pupil will be able to find the quotient and the remainder, if there is one.

Remaining Sums

Grade Levels: 4 - 6

Number of Participants: 2 - 4

Materials Required: Gameboard, 2 number cubes (dice covered with masking tape and the numerals 0, 1, 2, 3, 4, and 5), markers, paper, and pencil

483	2483	79	291	8465	1000
2010	213	43	357	5222	4876
479	3576	894	1010	3333	4973
1111	2242	8873	5034	786	99

Directions:

1. Each player rolls the die. The player rolling the highest number goes first.

2. The first player tosses a marker on the gameboard. If it falls on a line, the player chooses the adjacent number he or she wants to play.

3. The player then rolls the number cubes, adds the two numbers, and divides their sum into the number on which her or his marker landed.

4. After the player completes the division process, the remainder is added to his or her score.

5. Players take turns.

6. The first player to have remainders adding to 50 wins the game.

Wizard Division

Grade Levels: 5 - 6

Number of Participants: any number

Materials Required: Series of 4 different counters

Directions:

1. Mr. Wizard was given the problem 7)1435 to divide, but was given no paper or pencil.

2. He found the answer using four different sizes of rocks.

3. See if you can work the problem the same way Mr. Wizard did.

   Example:

   a. Mr. Wizard made 4 groups of stones with the largest (1 stone) standing for 1 thousand, the next largest (4 stones) representing 4 hundreds, the next (3 stones) for 3 tens, and the smallest stones (5 stones) for 5 ones. (See drawing A.)

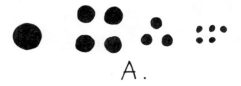

A.

   b. Mr. Wizard saw that there were not enough large stones to divide into groups of seven, so he traded the thousands stone for 10 stones representing hundreds. (See drawing B.)

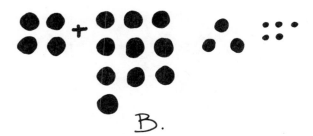

B.

   c. Now he was able to make 2 groups of 7 from the hundreds stones.

   d. He still could not divide the 3 rocks representing tens into groups of 7, so he traded these rocks for 30 rocks representing ones. (See drawing C.)

e. He was then able to make 5 groups of 7 from the smallest rocks.

f. As he looked at his solution he thought, "No group of thousands, 2 groups of hundreds, no group of tens, and 5 groups of ones. Two hundred and five is my answer." (See drawing D.)

g. Divide 7 ⟌1134 this way.

Winner's Circle

Grade Levels: 4 - 6

Number of Participants: 2 - 4

Materials Required: Gameboard, marker for each player, a die, and
3 groups of cards:
red cards contain basic division facts,
blue cards contain 3 digit dividends,
white cards contain 4 digit dividends.

red	safe	red	blue	white	blue	safe	white
white							red
white		white	go back 5 spaces	white	red		blue
safe		red			go back 6 spaces		red
blue		blue			Winner's Circle		white
red		safe					go back 2 spaces
start		white	white	blue	red	blue	white

Directions:

1. Each player rolls the die. The player rolling the highest number goes first.

2. The first player rolls the die and moves his or her marker that number of spaces on the playing board.

3. If the player lands on a red square, the player to his or her left draws a red card and gives him or her the problem written on it. (Red cards contain the basic division facts; e.g., $72 \div 9 = 8$.)

4. If the player answers correctly, he or she stays; if not, she or he must return to the last position.

5. If the player lands on a blue square, the player to her or his left draws a blue card for the player to answer. (Blue cards have 3 digit dividends; e.g., $287 \div 7 = 41$.) If correct, the player moves 1 extra space.

6. If the player lands on a white square, the player to his or her left draws a white card for the player to answer. (White cards have 4 digit dividends; e.g., $1458 \div 6 = 243$.) If correct, the player moves 2 extra spaces.

7. Players take turns.

8. The winner is the first player to reach the Winner's Circle.

Given any 4 digit numeral and a 2 digit divisor, the pupil will be able to find the quotient and the remainder, if there is one.

 # *Puzzle*

Grade Levels: 5 - 6

Number of Participants: any number

Materials Required: Spirit duplicating master

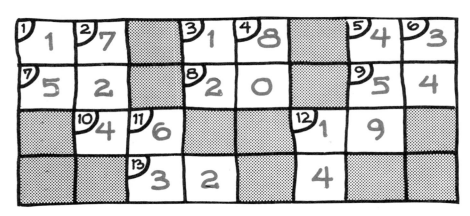

1. Solve the cross number puzzle by placing your quotient in the correct space.

Across	Down
1. 918 ÷ 54 = *17*	1. 1320 ÷ 88 = *15*
3. 1134 ÷ 63 = *18*	2. 2896 ÷ 4 = *724*
5. 817 ÷ 19 = *43*	3. 864 ÷ 72 = *12*
7. 4212 ÷ 81 = *52*	4. 3600 ÷ 45 = *80*
8. 1940 ÷ 97 = *20*	5. 1377 ÷ 3 = *459*
9. 1134 ÷ 21 = *54*	6. 1156 ÷ 34 = *34*
10. 1656 ÷ 36 = *46*	11. 1449 ÷ 23 = *63*
12. 361 ÷ 19 = *19*	12. 1386 ÷ 99 = *14*
13. 2816 ÷ 88 = *32*	

Batting a Thousand

Grade Levels: 5 - 6

Number of Participants: any number

Materials Required: A collection of newspapers showing baseball box scores. (Pupils will use the number of hits and "at bats" for certain ballplayers each day.)

Directions:

1. Players pick two of their favorite baseball players.

2. They chart the ballplayer's batting average daily. (The box scores of the previous day's games will be in the newspaper.)

3. Each player can record the number of hits and the number of official "at bats" for each ballplayer.

4. By dividing the total number of official "at bats" into the number of hits, the player can keep the batting averages. (If the players will chart each day's averages, they will get practice in the construction and use of graphs.)

5. Periodically the newspaper will show the batting average so that computation can be checked.

   Example: Pete Rose has 23 hits in 62 "at bats".

$$
\begin{array}{r}
.371 \\
62\overline{)23.000} \\
18\,6 \\
\overline{4\,40} \\
4\,34 \\
\overline{60}
\end{array}
$$
rounds off to .371 batting average

95

Divisor

Grade Levels: 5 - 8

Number of Participants: any number

Materials Required: Spirit duplicating master

Directions:

1. Fill in all of the blanks with the appropriate numerals so that the division steps are complete.

```
37)1258      83)3071      67)3149
  1110│30      2490│30      2680│40
   148│         581│         469│
   148│ 4       581│ 7       469│ 7
     0│34         0│37         0│47
```

```
73)1387      54)2268      27)2322
  730│10      2160│40      2160│80
  657│         108│         162│
  657│ 9       108│ 2       162│ 6
    0│19         0│42         0│86
```

Try this short cut for problems with zeros.

$80 \div 20 = \underline{8} \div \underline{2} = \rule{1cm}{0.4pt}$	$720 \div 90 = \underline{72} \div \underline{9} = \rule{1cm}{0.4pt}$
$630 \div 90 = \underline{63} \div \underline{9} = \rule{1cm}{0.4pt}$	$2700 \div 30 = \underline{270} \div \underline{} = \rule{1cm}{0.4pt}$
$490 \div 70 = \underline{} \div \underline{} = \rule{1cm}{0.4pt}$	$4500 \div 500 = \underline{} \div \underline{5} = \rule{1cm}{0.4pt}$
$540 \div 60 = \underline{} \div \underline{} = \rule{1cm}{0.4pt}$	$6300 \div 90 = \underline{630} \div \underline{} = \rule{1cm}{0.4pt}$
$720 \div 80 = \underline{} \div \underline{} = \rule{1cm}{0.4pt}$	$5600 \div 70 = \underline{} \div \underline{} = \rule{1cm}{0.4pt}$

Decode Division

Grade Levels: 5 - 6

Number of Participants: any number

Materials Required: Spirit duplicating master

Directions:

1. Decode the scramble of letters by working the problems.

2. Write the letter whose number matches the quotient of the problem in the space.

$\underline{M}$	$\underline{A}$	$\underline{T}$	$\underline{H}$		$\underline{C}$	$\underline{A}$	$\underline{N}$		$\underline{B}$	$\underline{E}$		$\underline{F}$	$\underline{U}$	$\underline{N}$.
1	2	3	4		5	6	7		8	9		10	11	12

1. $3510 \div 78 = 45$

2. $2494 \div 43 = 58$

3. $3358 \div 73 = 46$

4. $2077 \div 67 = 31$

5. $2744 \div 49 = 56$

6. $3848 \div 52 = 74$

7. $1960 \div 56 = 35$

8. $4368 \div 56 = 78$

9. $3915 \div 45 = 87$

10. $2184 \div 42 = 52$

11. $1113 \div 53 = 21$

12. $1116 \div 12 = 93$

Given a division problem where a zero is needed as a place holder in the answer, the pupil will be able to solve the problem correctly.

Verify

Grade Levels: 5 - 6

Number of Participants: 2

Materials Required: Cards with problems involving one factor in which zero is the middle digit, paper, and pencil

Directions:

1. Shuffle the cards and place the deck face down.

2. One player draws a card then writes the problem and answer for the other player to see.

3. The other player has 15 seconds to decide if the answer given is correct or incorrect.

4. If he or she decides correctly, the player keeps the card and gets another turn. If the player decides incorrectly, the card is returned to the bottom of the pile then he or she gives a problem to the other player.

5. The player who keeps the greatest number of cards wins.

6. Make cards as follows using the problems and answers given on the following page:

```
1442÷7=26

W 206
```

```
3216÷8=402

r
```

r 972 ÷ 9 = 108   303 w 2121 ÷ 7 = 33   307 w 3377 ÷ 11 = 37
· 503 w 4024 ÷ 8 = 54   r 915 ÷ 3 = 305   r 2427 ÷ 3 = 809
203 w 1827 ÷ 9 = 23   404 w 3232 ÷ 8 = 44   r 2828 ÷ 7 = 404
104 w 728 ÷ 7 = 14   r 2828 ÷ 7 = 404   r 7272 ÷ 9 = 808
306 w 1224 ÷ 4 = 36   701 w 6309 ÷ 9 = 71   r 9792 ÷ 102 = 96
r 1836 ÷ 6 = 306   503 w 3521 ÷ 7 = 53   r 7878 ÷ 78 = 101
202 w 1616 ÷ 8 = 22   r 2430 ÷ 6 = 405   r 918 ÷ 9 = 102
r 2412 ÷ 4 = 603   r 5454 ÷ 9 = 606   r 9078 ÷ 89 = 102
208 w 1664 ÷ 8 = 28   309 w 1854 ÷ 6 = 39

NOTE: Each card that has a "w" on it should have the correct answer next to the "w".

Boxed In

Grade Levels: 5 - 6

Number of Participants: any number

Materials Required: Spirit duplicating master

Directions:

1. Write the correct digit in the square to make the division problem correct.

Number "Shapes"

Grade Levels: 5 - 6

Number of Participants: any number

Materials Required: None

Directions:

a. A number is divisible by 2, if the number ends in an even number.

b. A number is divisible by 3, if the sum of the numerals is divisible by 3.

   Example: 4572   4 + 5 + 7 + 2 = 18 which is divisible by 3.

c. A number is divisible by 4, if the last two numerals are divisible by 4.

d. A number is divisible by 5, if the numeral ends in a 0 or a 5.

e. A number is divisible by 6, if it ends in an even numeral and if the sum of the numerals is divisible by 3.

1. If the numeral is divisible by 2, draw a square around it.

2. If it is divisible by 3, draw a right parenthesis.

3. If it is divisible by 4, draw a left parenthesis.

4. If it is divisible by 5, underscore it.

5. If it is divisible by 6, draw a rainbow over the top.

Example: 120 is divisible by 2, 3, 4, 5, and 6. It would look like this:

6. Test these numbers:

Division Journey

Grade Levels: 5 - 6

Number of Participants: 2 - 4

Materials Required: Gameboard, number cube (die covered with masking tape containing the numerals 2, 2, 3, 4, 5 and 6), 1 marker for each player, card with divisibility rules

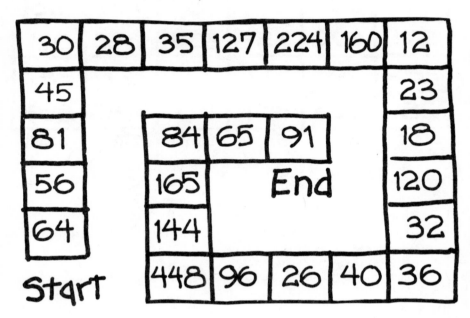

| 30 | 28 | 35 | 127 | 224 | 160 | 12 |

Directions:

1. Each player rolls the die. The player rolling the highest number goes first.

2. First player rolls the die and moves along the board stopping at the first number that is divisible by the number he or she rolled.

3. If the player skips a numeral that is divisible by the number rolled, the player must go back to the beginning. Play is checked by a student with divisibility rule card.

4. Players takes turns.

5. The first player who gets to the end is the winner.

To the Top

Grade Levels: 5 - 6

Number of Participants: any number

Materials Required: Spirit duplicating master

767	297	221	187	708	361
671	672	558	332	633	871
169	793	517	371	804	847
289	833	351	147	465	869
Start	Start	Start	Start	Start	Start

Directions:

1. Trace a path from one of the starting positions to the top of the chart by finding numbers that are divisible by any of the following numbers: 2, 3, 4, 5, or 6.

2. For each number used write the factor and divisibility rule used.

IDEA: Students could use a card with rules to help them find correct path.

Given a set of divisibility rules for 7, 9, 10, 11, and 13, the pupil will be able to apply the rules.

Quik Divisor

Grade Levels: 5 - 6

Number of Participants: any number

Materials Required: Paper and pencil

Directions:

I. A number is divisible by 7, if, when you strike out the last numeral, double that number, and subtract that number from the remaining digits, the answer is divisible by 7. (You may repeat this striking out and subtracting process until you obtain a number less than 100.)

Example:

    A. Is 343 divisibly by 7?

    B. Strike out the last numeral, 3. - - - 34$\cancel{3}$.

    C. Double the 3 and subtract from remaining digit.

$$\begin{array}{r} 34 \\ -6 \\ \hline 28 \end{array}$$

    D. Since 28 is divisible by 7, 343 is!

II. A number is divisible by 9, if the sum of the numbers is divisible by 9.

    Example: 864   8 + 6 + 4 = 18 which is divisible by 9, so 864 is divisible by 9.

III. A number is divisible by 10, if it ends in zero.

IV. A number is divisible by 11, if, when you strike out the last numeral and subtract that numeral from the remaining digits, the numeral left is divisible by 11. (You may repeat this striking out and subtracting process until you obtain a number less than 100.)

Example:

A. Is 869 divisible by 11?

B. Strike out the numeral, 9. - - - 86~~9~~.

C. Subtract the 9 from remaining numerals

$$\begin{array}{r} 86 \\ - 9 \\ \hline 77 \end{array}$$

D. Since 77 is divisible by 11, 869 is!

NOTE: The numbers you are subtracting in the case of 7 and 11 are really multiples of the prime number. This procedure seems to work for all of the prime numbers except 2 and 5.

1. Match the number on the left with a number it is divisible by.

2. Write the correct letter in the blank.

a. 438	_d_ 1. (9)
b. 649	_a_ 2. (6)
c. 4050	_b_ 3. (11)
d. 162	_e_ 4. (7)
e. 833	_c_ 5. (10)

V. A number is divisible by 13, if, when you strike out the last numeral, multiply that numeral by 9 and subtract this product from the remaining numerals, the numeral obtained is divisible by 13. (You may repeat this striking out and subtracting process until you obtain a number less than 100.)

Example:

A. Is 962 divisible by 13?

B. Strike out the numeral 2. - - - 96~~2~~.

C. Multiply the 2 by 9 and subtract from the remaining numerals.

$$\begin{array}{r} 96 \\ - 18 \\ \hline 78 \end{array}$$

D. Since 78 is divisible by thirteen, 962 is!

Come Eleven

Grade Levels: 5 - 6

Number of Participants: any number

Materials Required: Paper and pencil

Directions:

1. When using very large numbers, a good divisibility check for 11 is to add all of the digits in the odd numbered places starting at the right; i.e., ones, hundreds, and ten thousands place.

2. Add the digits in the even numbered places; i.e., tens, thousands, and ten thousands place.

3. Find the difference of the two sums.

4. If the difference is 0, 11, or 22, or any other multiple of 11, the number is divisible by 11.

   Example: Is 5,024,679 divisible by eleven? Adding odd place value digits from the right you get, 9 + 6 + 2 + 5 = 22. Adding even digits from the right you get 7 + 4 + 0 = 11. The difference between 22 − 11 is 11, so 5,024,679 is divisible by 11.

5. Try these numbers. Circle the ones that are divisible by 11.

47,835   5,808,154   3,461,926,453

215,371,954   4,091,516

10,987,647   1,624,050   9,209,167

Rule Memory

Number of Participants: any number

Materials Required: None

Directions:

1. Test your divisibility knowledge by determining which of the divisibility rules always applies when you:

   a. Multiply any chosen number by 7.

   b. Divide the product by 50. Find the quotient and the remainder.

   c. Add the quotient and remainder. This sum is always divisible by _7?_

Given a division problem, the pupil will be able to identify the place value of the first digit of the quotient.

L-o-n-g Division

Grade Levels: 5 - 6

Number of Participants: any number

Materials Required: Paper and pencil

Directions:

1. Given a problem such as 27)36749, the student will write the problem as:

27	10,000	1000	100	10	1 1 1
	10,000	1000	100	10	1 1 1
	10,000	1000	100	10	1 1 1
		1000	100	10	
		1000	100		
		1000	100		
			100		

2. The student then asks, the question, "How many groups of twenty-seven 10,000's can I form?" Since the answer is none, he or she renames each of the 10,000's as ten 1,000's.

27	1000	1000	1000	1000
	1000	1000	1000	1000
	1000	1000	1000	1000
	1000	1000	1000	1000
	1000	1000	1000	1000
	1000	1000	1000	1000
	1000	1000	1000	
	1000	1000		
	1000	1000		
	1000	1000		

3. The student repeats the question, "How many groups of twenty-seven 1,000's can I form?" He or she will determine that 1 group of twenty-seven 1,000's, can be formed. Then he or she marks that number about the 1,000's group.

4. This process will continue until the problem is solved.

Place It

Grade Levels: 5 - 6

Number of Participants: any number

Materials Required: Paper and pencil

Directions:

1. Given a problem such as 36)5472, rewrite it in the form:

$$\frac{54}{36} \bigg| 72$$

2. By examining the problem, it can be determined that 36 can be subtracted from 54, so the place value of the first digit will be in the hundreds column.

3. In the problem 71)1946, the rewritten form

$$\frac{194}{71} \bigg| 6$$

shows that 71 cannot be subtracted from 1, or 19, but only from 194 so the place value of the first digit will be in the tens column.

4. Using the same method in the following problems, determine in what column the first digit should be placed.

a. 36)1432   c. 29)3678

b. 61)967   d. 47)1679

108

Boxed Quotients

Grade Levels: 5 - 6

Number of Participants: 4, 8, 12, 16, 20

Materials Required: None

Directions:

1. Form teams of 4 players. Teams should be of equal ability.

2. The first player of each team writes a division problem, dictated by the teacher, on the chalkboard.

3. This player draws a box above the appropriate place value position for the first number in the quotient.

4. The next player writes the appropriate digit in the box then performs the corresponding multiplication and subtraction.

5. The next player does the work of the next place value digit.

6. If a mistake is made in solving the problem, the next player must start from the beginning.

7. The first team that completes the problem successfully wins.

Example:

1st player

$$17\overline{)11084}$$
with an empty box above

2nd player

$$17\overline{)11084}$$
$$10200$$
$$884$$
with a boxed 6 above

3rd player

$$17\overline{)11084}$$
$$10200$$
$$884$$
$$850$$
$$34$$
with boxed 6, 5 above

4th player

$$17\overline{)11084}$$
$$10200$$
$$884$$
$$850$$
$$34$$
$$34$$
with boxed 6, 52 above

Given a division problem, the pupil will be able to find the quotient and then identify a valid procedure for checking the answer.

Relationships

Grade Levels: 4 - 6

Number of Participants: any number

Materials Required: Spirit duplicating master

Products of two factors	factor or addend	factor or addend	sums of two addends
99	11	9	20
36	18	2	20
50	5	10	15
55	5	11	16
56	8	7	15
9	1	9	10
60	15	4	19
45	15	3	18
60	5	12	17

Directions:

1. Fill in the blanks on the chart.

2. Fill in the blanks of the sentences:

   a. If you are given a product and a factor, you can find the other factor by *dividing*.

   b. Since division is the inverse of *multiplication*, we can check our division by *multiplying* the factors to obtain the product.

Up

Grade Levels: 4 - 6

Number of Participants: any number

Materials Required: Spirit duplicating master

quotient	product	factor	equation
7	56	8	7×8 =56
15	225	15	15×15=225
9	81	9	9×9=81
13	156	12	12×13=156
56	2408	43	43×56=2408
53	3763	71	71×53= 3763
51	3213	63	51×63 = 3213

Directions:

1. Fill in the blanks.

Add-Em-Up

Grade Levels: 5 - 6

Number of Participants: any number

Materials Required: Paper and pencil

Directions:

1. Any division problem can be checked in the following manner. Let's use the following problem as an example:

$$24\overline{)8664} = 361$$

a. Add the digits in the quotient $(3 + 6 + 1 = 10)$. Keep on adding the digits until a one digit number is reached, (10 is a two digit number so add $1 + 0 = 1$).

b. Do the same for your divisor $(2 + 4 = 6)$. Six is the number to remember for your divisor.

c. Multiply the single digit quotient by the single digit divisor $(1 \times 6 = 6)$. If your answer is correct, when you add the digits in your dividend, your answer should also be 6.

   NOTE: When you multiply the one digit quotient by the one digit divisor and get a two digit answer keep adding until you get a one digit number.

d. Now add the digits in the dividend $(8 + 6 + 6 + 4 = 24)$. Your answer is a two digit number so keep adding the digits until a one digit number remains $(2 + 4 = 6)$.

   Your answer is correct.

$$24\overline{)8664} = 361$$

   Quotient X Divisor = 6
   Dividend = 6

   NOTE: If the quotient has a remainder, add the digits in the remainder until you have a one digit remainder. Then, add the remainder to the one digit number you get when you multiply the one digit quotient by the 1 digit divisor.

Given a number sentence involving multiplication and division, the pupil will be able to select the one that shows the correct inverse relationship.

A Close Knit Family

Grade Levels: 4 - 6

Number of Participants: any number

Materials Required: Spirit duplicating master

	$2+3=$ _____ $6 \times$ _____ $= 30$ $3 \times$ _____ $= 6$ $5 = 6-$ _____	$3 \times 2 = 6$ $2 \times 3 = 6$ $6 \div 3 = 2$ $6 \div 2 = 3$
give your very best	$1 \times 4 =$ _____ $24 \div 6 =$ _____ $4 \times$ _____ $= 16$ $4 +$ _____ $= 8$	$4 \times 4 = 16$ $16 \div 4 = 4$
	$21 \div 3 =$ _____ $25 - 21 =$ _____ $4 \times$ _____ $= 20$ $21 - 7 =$ _____	$21 \div 3 = 7$ $21 \div 7 = 3$ $3 \times 7 = 21$ $7 \times 3 = 21$
	$3 \times 4 =$ _____ $24 \div 6 =$ _____ $4 \times$ _____ $= 16$ $4 +$ _____ $= 8$	$3 \times 4 = 12$ $4 \times 3 = 12$ $12 \div 3 = 4$ $12 \div 4 = 3$
	$25 \div 25 =$ _____ $2 \times 25 =$ _____ $25 +$ _____ $= 50$ $5 \times$ _____ $= 25$	$5 \times 5 = 25$ $25 \div 5 = 5$

Directions:

1. Mark the combination suggested by the picture.

2. Write the entire fact family.

Fill 'Er In

Grade Levels: 4 - 6

Number of Participants: any number

Materials Required: Spirit duplicating master

77	96	72	108	75	132	35	45	91	56	16	64	49	54
7	12	9	12	15	11	7	9	13	7	4	8	7	9
11	8	8	9	5	12	5	5	7	8	4	8	7	6

Directions:

1. Fill in the blank spaces when you have discovered the pattern.

Genealogy Search

Grade Levels: 4 - 6

Number of Participants: any number

Materials Required: Spirit duplicating master

Multiplication Facts		Division Facts	
7×8 = 56	8×7=56	56÷7=8	56÷8=7
9×5 = 45	5×9=45	45÷9=5	45÷5=9
9×7 = 63	7×9=63	63÷7=9	63÷9=7
6×9 = 54	9×6=54	54÷9=6	54÷6=9
7×6 = 42	6×7=42	42÷7=6	42÷6=7
5×7 = 35	7×5=35	35÷7=5	35÷5=7
4×9=36	9×4=36	36÷4=9	36÷9=4
8×9=72	9×8=72	72÷9=8	72÷8=9
6×8=48	8×6=48	48÷8=6	48÷6=8

Directions:

1. By knowing one multiplication or division fact, you can search out three other facts.

2. Fill in the chart. The first one is done for you.

Equation War

Grade Levels: 4 - 6

Number of Participants: played by pairs of students

Materials Required: Spirit duplicating master

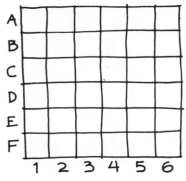 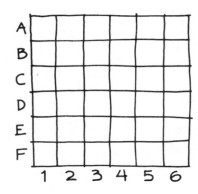

Directions:

1. Each player hides 5 division equations on one of the grids.

2. The other grid is used to record the other player's resources.

3. One of the players gives a location on the grid by calling a letter and a number. (Example: B,5 would be the location of 36, part of one of opponents division equations.)

4. If a hit is scored, the opponent must tell what numeral or symbol was hit. (For example, opponent would identify B,5 as being 36.)

5. If the player can state the full equation correctly, he or she gets to claim the full equation for his or her turn.

6. The first player to name all of his or her opponents equations wins the game.

115

Given up to 3 numbers, the pupil will be able to find the greatest common factor.

GCF Race

Grade Levels: 6

Number of Participants: any number

Materials Required: Gameboard and 25 cards (2, 2, 3, 3, 4, 4, 5, 6, 7, 8, 9, 10, 12, 14, 15, 16, 18, 20, 21, 24, 27, 30, 32, 36, 40)

1	2	3	4	5	6	7	8	9	10
20	19	18	17	16	15	14	13	12	11
21	22	23	24	25	26	27	28	29	30
40	39	38	37	36	35	34	33	32	31
41	42	43	44	45	46	47	48	49	50
60	59	58	57	56	55	54	53	52	51
61	62	63	64	65	66	67	68	69	70
80	79	78	77	76	75	74	73	72	71
81	82	83	84	85	86	87	88	89	90
100	99	98	97	96	95	94	93	92	91

Directions:

1. Each player draws one card. The player with the highest card goes first.

2. Return all cards to the deck and shuffle. Each player draws 2 or 3 cards (depending on ability of player) at the beginning of his or her turn to play.

3. Each player will determine the GCF of his or her cards.

4. The first player moves as many spaces as his or her GCF.

5. All players take a turn.

6. At the conclusion of round one all cards are returned to the deck. The deck is reshuffled and steps 2 through 5 are repeated.

7. First player who arrives at 100 wins.

Factor Puzzle

Grade Levels: 6

Number of Participants: any number

Materials Required: Spirit duplicating master

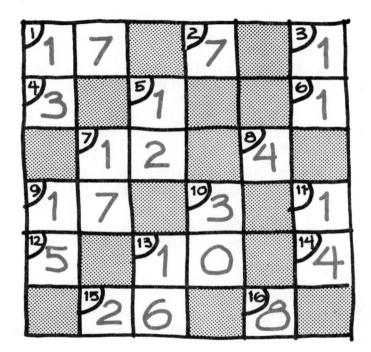

Directions:

1. Work the following GCF cross number puzzle.

Across		Down
1. 17, 34, 51	9. 34, 51	1. 26, 39, 52
2. 14, 28, 35	10. 12, 27, 9	3. 22, 44, 77
3. 9, 17, 23	11. 13, 21, 7	5. 36, 48, 60
4. 15, 21, 36	12. 35, 55, 60	7. 17, 51
5. 4, 9	13. 20, 30, 40	9. 30, 75
6. 7, 8	14. 8, 16, 28	10. 60, 150
7. 24, 48, 60	15. 26, 52	11. 28, 42
8. 12, 16, 20	16. 16, 24	13. 32, 48
		16. 24, 32, 48

Euclid's Maze

Grade Levels: 6

Number of Participants: any number

Materials Required: Lots of brain power!

Directions:

1. To determine the greatest common factor (GCF) of two large numbers, divide the smaller number into the larger.

2. The remainder obtained becomes the new divisor and the dividend of the first division problem becomes the new dividend. Divide the dividend by the divisor.

3. Repeat the process until the remainder obtained is zero.

4. The last divisor used is the GCF of the two large numbers.

Example: Find the GCF of 952 and 731.

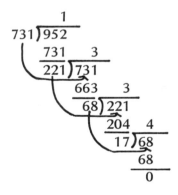

Therefore, 17 is the greatest common factor.

Example: Find the GCF of 301 and 329.

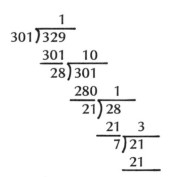

Therefore, 7 is the greatest common factor.

118

Using this procedure determine the answer to the following problem.

1. Choose any number.

2. Multiply that number by the GCF of 111 and 171. *(3)*

3. Add the GCF of 143 and 221. *(13)*

4. Subtract the GCF of 90 and 110. *(10)*

5. Divide by the GCF of 249 and 201. *(3)*

6. Subtract the number you first selected.

THAT NUMBER IS WHAT YOU ARE!

Appendixes A and B

Basic Multiplication Skills-Sample Test

Information for the User: The number beside each problem corresponds to the number of the objective which that problem represents. It is suggested that the problem selected for use be written on a spirit duplicating master. Direct the students to circle or write each correct answer.

1. Circle the counters that illustrate this statement:

   3 sets of 4 are 12.

   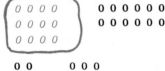

   ```
   ( 0 0 0 0 )      0 0 0 0 0 0
   ( 0 0 0 0 )      0 0 0 0 0 0
   ( 0 0 0 0 )
   ```

   ```
   0 0        0 0 0
   0 0        0 0 0
   0 0        0 0 0
   0 0        0 0 0
   0 0
   0 0
   ```

2. Write the repeated addition expression and its sum for the following picture.

   ● ● ●
   ● ● ● $2 + 2 + 2 = 6$

3.
   $$\begin{array}{r} 7 \\ \times\,0 \\ \hline 0 \end{array} \qquad \begin{array}{r} 8 \\ \times\,1 \\ \hline 8 \end{array}$$

4.   5.
   $$\begin{array}{r} 4 \\ \times\,5 \\ \hline 20 \end{array} \qquad \begin{array}{r} 9 \\ \times\,6 \\ \hline 54 \end{array}$$

6.   7.
   $$\begin{array}{r} 13 \\ \times\,3 \\ \hline 39 \end{array} \qquad \begin{array}{r} 18 \\ \times\,4 \\ \hline 72 \end{array}$$

8.
   $$\begin{array}{r} 64 \\ \times\,17 \\ \hline 448 \\ 64 \\ \hline 1088 \end{array}$$

9. What is the product of these factors: 56, 15.

   $$\begin{array}{r} 56 \\ \times\,15 \\ \hline 280 \\ 56 \\ \hline 840 \end{array} \qquad \begin{array}{r} 15 \\ \times\,56 \\ \hline 90 \\ 75 \\ \hline 840 \end{array}$$

10. Group the following factors in two different ways. Find the product of each.

    3, 6, 2

    Possible answers:

    $3 \times 2 \times 6 = 36$
    $6 \times 3 \times 2 = 36$

11. Circle the expression that has the same value as 8 x 14.

    $(6 \times 12) + (2 \times 2)$

    $(5 \times 14) + (3 \times 14)$

    $(6 \times 7) + (2 \times 7)$

12. $100 \times 37 = 3700$

13.
   $$\begin{array}{r} 212 \\ \times\,149 \\ \hline 1908 \\ 848 \\ 212 \\ \hline 31588 \end{array}$$

14. Circle the factors for 30.

    5, 6   2, 7   3, 8

15. What is the complete factorization for 16?

    $2 \times 2 \times 2 \times 2$

16. What is the least common multiple for 2, 3, 4? 12

17. $30 \times 8000 = 240000$

18.
   $$\begin{array}{r} 234 \\ \times\,55 \\ \hline 1170 \\ 1170 \\ \hline 12870 \end{array}$$

Basic Division Skills-Sample Test

Information for the User: The number beside each problem corresponds to the number of the objective which that problem represents. It is suggested that the problems selected for use be written on a spirit duplicating master. Direct the students to circle or write each correct answer.

1. Partition this set into 5 equivalent sets.

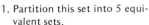

2. How many sets of 3 objects are in this set?

Answer: *6*

3. $21 \div 3 = $ *7*

4. $40 \div 5 = $ *8*

5. $72 \div 8 = $ *9*

6.
$$6 \overline{)1026} \quad ^{171}$$

7.
$$23 \overline{)1449} \quad ^{63}$$
$$\underline{138}$$
$$69$$
$$\underline{69}$$

8.
$$24 \overline{)4848} \quad ^{202}$$

9. Write the number (or numbers) that is in the parenthesis on the left beside the number (or numbers) on the right that it is divisible by. A number on the right may be divisible by more than one number on the left.

(2)	a. 655	*(5)*
(3)	b. 762	*(2, 3, 6)*
(4)	c. 314	*(2)*
(5)	d. 248	*(2, 4)*
(6)	e. 723	*(3)*

10. Write the number (or numbers) that is in the parentheses on the left beside the number (or numbers) on the right that it is divisible by. A number on the right may be divisible by more than one number on the left.

(7)	a. 693	*(7, 9)*
(9)	b. 371	*(7)*
(10)	c. 830	*(10)*
(11)	d. 441	*(7, 9)*
(13)	e. 572	*(11)*

11. What is the place value of the first digit of the quotient for

$$47 \overline{)1679} \quad ^{?}$$

tens place

12. Find the quotient and check your answer:

$$14 \overline{)196} \quad ^{14}$$
$$\underline{14}$$
$$56$$
$$\underline{56}$$

Check
$$14$$
$$\underline{\times 14}$$
$$56$$
$$\underline{14}$$
$$196$$

13. Write the entire fact family for $36 \div 9 = 4$

36 ÷ 4 = 9
9 x 4 = 36
4 x 9 = 36

14. What is the greatest common factor for 14, 28, 35?

Answer: *7*

Multiplication Profile Form

Suggestions for recording results: √ has little or no understanding; ⊘ has some understanding, but lacks mastery; ● answered problem correctly.	Multiplication Objectives					
Pupil	1.	2.	3.	4.	⟶	18.
1.						
2.						
3.						
4.						
↓						
35.						

Division Profile Form

Suggestions for recording results: √ has little or no understanding; ⊘ has some understanding, but lacks mastery; ● answered problem correctly.	Division Objectives					
Pupil	1.	2.	3.	4.	⟶	14.
1.						
2.						
3.						
4.						
↓						
35.						